AUDIOVISUAL EQUIPMENT AND MATERIALS:

A Basic Repair and Maintenance Manual

VOLUME II

by
Don Schroeder
and
Gary Lare

The Scarecrow Press, Inc.
Metuchen, N.J., & London   1989

British Library Cataloguing-in-Publication data available

Library of Congress Cataloging-in-Publication Data
(Revised for volume 2)

Schroeder, Don, 1931-
    Audiovisual equipment and materials.

    Includes bibliographical references.
    1. Audio-visual equipment--Maintenance and repair.
I. Lare, Gary
TS2301.A7S35   621.389'7   79-384
ISBN 0-8108-1206-1
ISBN 0-8108-2265-2 (alk. paper)

CONTENTS

FOREWORD

About two years after the publication of <u>Audiovisual Equipment and Materials: A Basic Repair and Maintenance Manual</u>, correspondence was begun relative to a second edition. That book had stopped at black and white, reel-to-reel video tape recorders (does anyone even remember VTRs?), with the whole color videocassette acceptance and widespread school use of microcomputers still to come. The hardware of educational technology had moved to much higher cost and complexity, usually requiring the skills of a professional serviceperson to restore function to defunct equipment.

On the other hand things hadn't changed all that much in those devices that were covered in <u>Audiovisual Equipment and Materials</u>.... Certainly newer models had been introduced. Others were discontinued. Some later adhesives and lubricants had not proven to be as great as claimed, and technicians were back to using what we had recommended. There were a few improvements in light efficiency and in microphones and speakers, but the discussions of basic principles had not changed. Some improvement of the organization of the book was possible, but it would be the same content in new guise.

Over the years, when problems would come up by phone or during school visits, they would often occasion checking with the text and pictures to see what had been said. Repeatedly we were surprised to find what we still considered to be adequate coverage. Though the book concentrated on "old" technology and projectors and audio record players, tape recorders and PA systems, it consistently said what we wanted to say about those devices. If it was inadequate, it was the publication date that imposed limitation in covering newer technologies, rather than in what was said about the equipment widely in use during the writing of the book in 1977 and 1978.

It is now more than ten years later. Video display has made a strong inroad on the projected image. Microcomputers are fulfilling the roles so long foretold by prophets, providing interactive teaching units and drill and practice, and enhancing many aspects of management of the classroom and measurement and recording of diagnostic and prescriptive test data.

Had it not been for concurrent improvement in the new basic electronic building blocks--semiconductor chips--and vastly improved reliability, this new world of educational technology would have become a service nightmare. But for all its incredible complexity--and it is complex almost beyond belief and certainly beyond what anyone watching the consumer electronics boom of the post World War II era would have envisioned--the attendant reliability is no less remarkable. Serious electronic problems within the devices are relatively rare, particularly during the first two or three years of use.

Prices are low enough that those units developing serious reliability problems can be written off and replaced, usually with a very low pro-rated cost for the service they have given. Still, there are things that personnel charged with managing this equipment can and should do, and know how to do, that will keep a program on the screen when things go wrong and that will extend the useful life of the equipment.

We believe there is sufficient information now available to merit a second volume, rather than a second edition. Neither of the authors is a skilled service technician, but both have saved enough presentations to realize that they can share knowledge that gives an edge over barely knowing what to do, particularly when it is clear that someone needs to do something. Our intended audience is still essentially the same: people working in various educational settings whose primary role is not electronics-technical, but who are charged with keeping some technically supported enterprise operational.

One would have thought, with the wide use of VCRs and computers at home, a greater expertise in coping with some of the simpler problems would have become common in society. On the contrary, it seems to have gotten worse. Unless the equipment "at school" is exactly like the equipment at home, confusion reigns. Many of us with whole campus or school district responsibilities are still running too many fool's errands to correct what should have been obvious to the person using the equipment, or to someone responsible at the "local" level.

And so we think there is a place for this book. Granted, it may often be less thorough than the instruction book that comes with a particular piece of equipment. And one reason for that is that we don't know just what kind of equipment is in your installation. But it does bring a lot of practical, in-service proven information into one reference book. This volume concentrates on the usual system cable problems--still an overwhelming source of difficulty. It discusses TV receivers and monitors, VCRs and CD and Video Disc units, and microcomputers. It also looks at some of the cable systems that tie it all together.

Unfortunately, we can probably not claim that it will save the money on repair costs that the earlier book could. But timely use of suggestions in this book may often save a program, and when the room is full of people, that's worth something.

This stuff is complicated and when it has real troubles they should be left to a service professional. Poking around inside a color TV or monitor with the power on is no place for a novice. The voltages are lethal, and the probability of actually fixing anything is negligible. But we work in people-intensive environments and too many of those people never keep their hands off of anything. We can help you determine the difference between real trouble and the kind that results from a casually twisted brightness or contrast control, or a bent wire in an F-connector. And even as this book is being written, solid-state video displays are becoming more common. Though their technology is also complex, they will remove the last of the dangerous voltages and make service attempts relatively safe, if not assured of success.

The authors would like to acknowledge the work of Lisa Sparks in developing the graphics for this book; the practical bench service experience shared by Bob Niemoeller, AV repair shop technician with the Cincinnati Public Schools; Lee Nourtsis and Danny Thomas for proving that two bright high school students can install belts and a head in a VHS VCR successfully.

Don Schroeder
Gary Lare

I.   SERVICE CONCEPTS

SERVICE CONCEPTS

SERVICE LOGIC

The key to diagnosis of problems with instructional equipment is application of a service logic, which is really the scientific method applied to equipment failure. All service work is done this way, either consciously or subconsciously. The steps are:

DEFINE THE PROBLEM, usually by asking or being told or shown what the equipment is, or is not, doing. If you are at all skeptical, confirm the problem yourself. This will reveal operator errors or controls and cables that have been left in some abnormal configuration by a previous user or inept fixer-upper. Having established that a technical problem exists ...

MAKE OBSERVATIONS, using all your senses, in an effort to localize the problem. This also requires a sound knowledge of "normal" operation, or in the case of cables, interconnection. If you do not have this knowledge, try to find the original operating instructions that came with the equipment. Try to get a clear idea in your mind of what is working and what is not. Then ...

TEST THE HYPOTHESIS by taking some kind of corrective action or making a further test or measurement. If your first guess isn't right, rejoice, you're still human. Make another hypothesis and test it. The making of hypotheses is creative work, and often amenable to a cup of coffee away from the problem, a walk around the block, or a night's fitful sleep. Having "tried every conceivable thing," we conceive yet another thing when we are away from the problem.

If your test shows you are correct, and you are able to do it, MAKE THE REPAIR.

If your tests all prove inconclusive, or you fear the repair is beyond you, SEND THE EQUIPMENT OUT FOR REPAIR.

SERVICE KNOWLEDGE

The knowledge of combined electronics and mechanics required to repair something like a videocassette recorder is unique. It's not something that just anyone can do, or even learn to do, and it is no put-down to refer such repairs to those who have that knowledge.

Furthermore, increasingly large amounts of specialized information is actually, or border-line, proprietary. Special service seminars are conducted by manufacturers with attendance limited to technicians who work for franchised dealers. Service manuals are often difficult to obtain, and service bulletins describing up-dates and modifications are distributed only to "authorized service agencies."

But that is all knowledge of what goes on inside the various units that comprise a system. The introduction of transistorized and integrated circuits with their generally lower operating voltages and temperatures has made possible incredible complexity, wondrous miniaturization, and with it all, fantastic reliability. (Annoying as it can be, it is almost amusing when it is the clock in a VCR that fails.)

Most institutional equipment problems stem from inexperienced operators, multiple users, and mobility. People make mistakes, get confused, or just don't know what they are doing. Most electro-mechanical equipment seems to work best when the number of different operators is restricted, to one if possible. Bumping along on carts over raised door thresholds and ex-pansion joints, over pot-holed streets in trucks, and through narrow doors snagging and bend-ing cables--these cause most institutional equipment problems, and correction rarely takes in-depth knowledge of the equipment.

In the discussion of observations under Service Logic, "a sound knowledge of normal oper-ation" was mentioned. This cannot be overstressed. How can you hope to find malfunction if you do not know proper operation when you see it? The best time to gain this knowledge is when the equipment is new and the instruction booklet is available. All new equipment should be checked for proper operation anyway, and that's a great time to try all the knobs, buttons, levers, switches, etc. The knowledge gained at this time can be useful in helping staff learn proper use, in running routine performance checks, and is essential when malfunction is reported.

A reputation for being able to fix things only requires that you have more knowledge and/or willingness to analyze problems than those who call for help. Nor is the helpful knowl-edge always technical. Sometimes it's knowing where a possible substitute unit can be found, and having the authority, or knowing how to get it, to make the substitution until a permanent remedy can be worked out.

Delivery of the program is the paramount objective. It is the reason most of our institu-tions exist, and the person whose knowledge makes that possible will be esteemed by colleagues and the organization served.

SERVICE SKILLS

The skill required to replace a VCR head assembly or belt set is generally higher than that necessary to repair a filmstrip projector. Not only does the VCR require skill in what to do--it requires skill to keep from doing as much or more harm than good. The slip of a screwdriver, scratching a polished tape drum or guide surface, could be more serious than the original problem. Excessive pressure to loosen a tight screw could cause a mis-alignment that would be very hard to correct.

Most of us can learn to replace a coaxial connector or to splice a power cable, even if soldering is involved. Removing and replacing a large scale integrated circuit on a circuit board, on the other hand, not only requires great skill and care, but special tools as well. Even integrated circuit chips that plug into a socket must be handled with understanding to assure that body static does not destroy the chip as it is inserted into the circuit. Some whole pieces of equipment may prove too delicate or complex for general institutional use, particularly as miniaturization for the consumer market dictates design, and skill required for use may become as much a problem as skill required for repair.

The ability to solder is a basic skill in electronics work. A small pencil type soldering iron works equally well for cable connectors and circuit boards. Many of them heat very quickly, which is a useful feature for the fast repairs often needed to restore operation of something even as an audience is coming into the room. Only experience can teach you how much to tighten, how hot to solder, how tight to grip. As with any other skill, practice is essential to improvement. Knowing when to give up also helps, because some things just have to be referred to more skilled hands.

AN EXAMPLE OF SERVICE LOGIC, KNOWLEDGE, AND SKILL

A complaint is made that a particular computer is not reading discs. The user further contends that the software is "all right" and has worked in other computers.

You first put the software in the original disk drive and try to load it into the computer to confirm the complaint (L). The program does not boot. The disk drive light does not even come on. Also, a message is displayed which says no drive exists. But you are looking right at the disk drive with a cable going from it into the computer. Since the computer is often moved, this can disturb the various connections (K). You hypothesize that the problem lies in the cable connections to the disk drive (L). You unplug the computer and remove its cover (K,S), to gain access to the disk drive connection to test the hypothesis (L). You find that the connection is indeed loose. You apply pressure to firmly seat the plug (S). Normal operation is restored.

A FEW WORDS ABOUT ELECTRICAL SAFETY

Most of the equipment discussed in this book uses 117 volt line power, and that which does not may use it to recharge the batteries. This rather high, potentially dangerous and lethal voltage and current is fortunately usually limited to the power cord, power switch, and any fuses within the equipment. But to get power into the power transformer, it may be routed via foil paths on the circuit board. For this reason it is always good practice to have the power unplugged at the wall before removing any covers and while handling the device. It is so easy to have a finger touch a "hot" part of the circuit board while turning a piece of equipment over to look at the other side.

Once line power gets to the isolating power transformer, most modern equipment will operate on voltages of around 12 to 20 volts ... not even enough to shock if touched. It does take an experienced eye to trace where the high line voltage is, and where the safer low voltages are. The transformer also isolates the operating voltage from "ground."

The exception to this is any Cathode Ray Tube device, such as a television receiver or monitor, or a TV camera. CRT devices, regardless of their low operating voltage, must have a section of very high voltage to make the cathode ray tube work. For this reason we strongly encourage referring any repair on this equipment to qualified and experienced service personnel.

When working with 117-volt line power, care must be taken regarding the electrical "ground." In power distribution in the United States this refers to a reference potential of 0 volts when measured to an earth ground (a metal rod driven about 4 feet into the earth). One wire of the power distribution system is actually grounded. The other wire, referred to as the "hot" wire, is then 117 volts "above ground." It is thus possible to get a lethal shock just by touching one wire (the hot one) when you stand on a wet floor in your bare feet. The metal rim of a sink top or plumbing fixture connected to a water pipe (ground) can also complete a shocking circuit.

If you are concerned about shock hazard, possible in an installation where electrical outlets are above a sink, or where you have novice service personnel, consider installing "ground fault interrupter" outlets. A duplex ground fault outlet costs about $20 and will shut off the power fast enough to prevent a lethal shock, should anything or anyone complete a circuit between ground and the hot wire. It's really cheap insurance against a shocking situation.

Most equipment is quite safe when functioning normally. But wires become

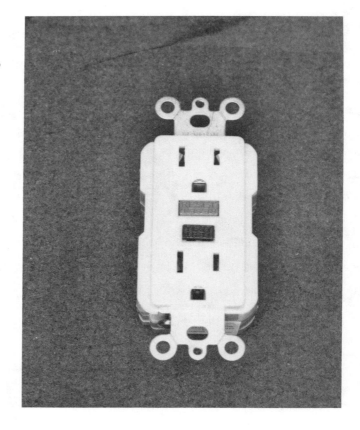

frayed with use, and careless or unskilled individuals sometimes make sloppy repairs, leaving tiny strands of wire touching metal frames or casings. To provide a measure of protection the three-wire power cord has been offered in recent years. The round prong of a three-prong plug should actually make a true ground connection between any metal frame or cabinet and an earth ground, independent of the grounding of the power distribution system. A circuit short within the equipment may then blow a fuse or circuit breaker, but it should stay safe for the operator. Use of two-prong adapters, without grounding the green wire or lug, subverts this whole protection system.

Do not rely on asphalt tile as an insulator from electrical shock. It would probably prevent lethal current levels, but it does pass some voltage when laid on concrete floors, especially on the ground level of buildings (where media departments so often seem to be located).

Take time to unplug equipment before removing the case. You can always plug it in again when you are past the danger of a finger slipping into an exposed terminal with line voltage on it. An "off" switch may turn the equipment off, but inside the case the line power is all there at the end of the cord. Since so many of our repairs are cord and plug replacements, take

time to do them right and carefully. If possible, try to wire the new wires to the same plug prongs as the original ones were, and if the cord is being replaced, wire white, black and green wires where the corresponding dolors of the original cord were. Before closing a plug or recasing a unit, inspect your work for any little loose strands of wire and cut off any you find.

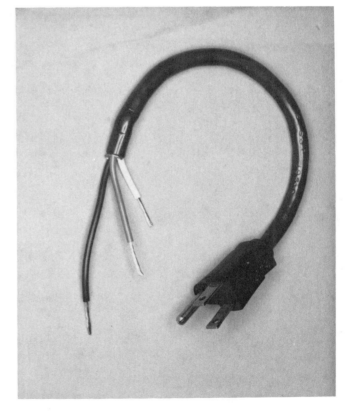

If you are unsure of these procedures, seek a demonstration from someone who does this kind of work. Anyone in this society should be able to replace a power plug, but we all have to learn how to do it right.

While specific use of tools such as a screwdriver for its intended purpose of removing or tightening screws, or as a pry tool to remove an integrated circuit chip is fine, don't just randomly go probing around in equipment or on circuit boards with screwdrivers, pieces of wire, or soldering aids. Unless you know what you are doing, this will amost invariably cause even more trouble and a higher repair bill than you would have had in the first place. You can also mess up the logic processes of a competent technician. Two unrelated problems simultaneously are very difficult to rationalize, and an unadmitted person-caused problem inside equipment is outside the logic chain. Experience at this level should be gained in courses or supervised apprenticeship, not by aimlessly poking around to see what might happen.

Remember, there are two levels of the electrical safety problem: your own, and all the rest of us. Since this book assumes equipment used in some sort of institutional setting, repair entails a degree of public responsibility beyond that of fixing your own home appliances. When we do the work ourselves, we don't want future charges of endangering others because of our lack of knowledge or skill. And if we have a supervisory role, it is our responsibility to see that people who take the work seriously are doing the repairs.

II.   EQUIPMENT MAINTENANCE AND REPAIR

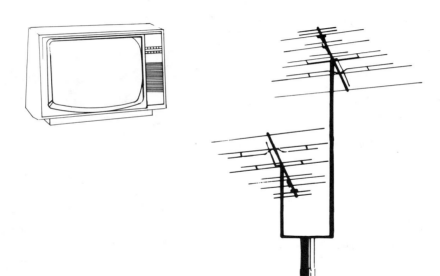

TELEVISION

BROADCAST TELEVISION IN THE INSTITUTIONAL SETTING

Reception of broadcast signals for use in classrooms and other institutional settings is usually not as easy as it is in the home. Because of the very high safety standard required by building codes, extensive use of metal and masonry is common. These materials seem to impede good signal penetration within the building, making use of various set-top rabbit-ear and wire loop UHF antennas a chancy proposition. Good color reception requires a signal that is strong and constant. "Snow" and fading or shifting color are almost always an indication of poor signal reception.

For years it has been recommended that TV receivers be placed along a window wall to reduce reflection of outside light off the glass screen surface. Shades, blinds or drapes are drawn behind the receiver to reduce ambient glare for those watching the program. This place-ment also has a reception advantage in that it moves any internal antenna near the windows, which may improve reception if the signal can get to the antenna without going through a lot of building material. In general, the UHF signals are more vulnerable to weakness, ghosts, etc. than VHF signals, although this could vary depending on strength of the transmitter, height of the transmitting antenna, and proximity of the transmitter to the receiving antenna. When an outside antenna is just not possible, consider the following:

For UHF:

- rigging a UHF antenna intended for outdoor use into or onto a microphone stand, standing it near a window, and rotating it for best picture.

- using an inside rabbit-ear antenna with a rotatable UHF pair of loops in the center. We have found those with a horizontal plastic ring around the two wire loops effective in difficult situations.

For VHF & UHF:

- establish a viewing room (like the old projection rooms) high in the building where the prospect of good signal reception is better.

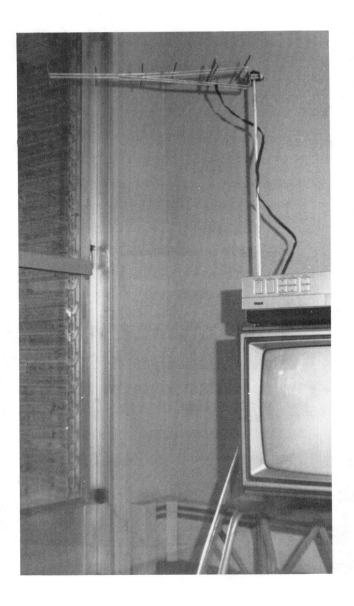

EXTERNAL ANTENNAS AND DISTRIBUTION SYSTEMS

If at all possible, an external antenna should be used. These come in varying degrees of complexity, usually dictated by the money available at the time of installation, and the needs of the institution.

Simplest of all is a broadband (all VHF & UHF channels) antenna with a lead-in wire coming down to a single area of use. If the desired stations are spread around the compass, a rotor may be included, with control at the TV receiver.

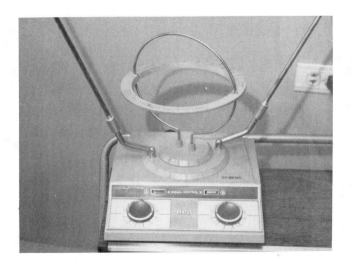

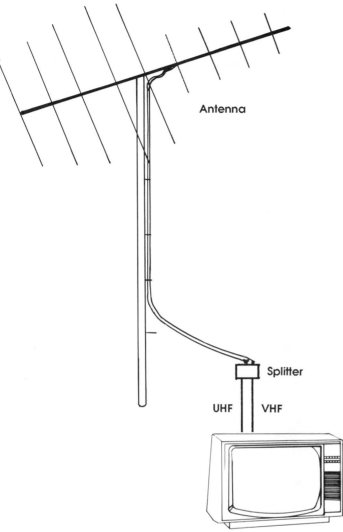

At the receiver a VHF/UHF splitter is used to provide appropriate TV signal to the separate connections on the receiver. The lead-in wire may be a single black coaxial cable with an F-connector, or it may be 300-ohm twin-lead.

There is a variety of splitters available. You have to figure out the combination you need to specify and purchase the right splitter.

The problem is that connections to the receivers vary. Equipment made for the institutional market will normally have a type F socket for VHF and two terminal screws for UHF.

You can combine splitters and 300-ohm to 72-ohm matching transformers, but remember that each device results in a slight loss of signal that can cost picture quality. In a simple system of this type, sudden loss of picture quality is usually the result of vandalism to the lead-in wire, corrosion or wire breakage up on the antenna, or high wind toppling or otherwise damaging the antenna.

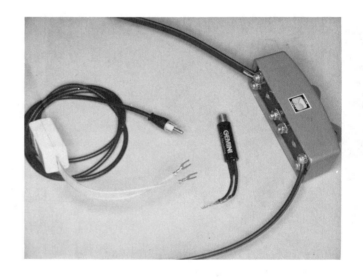

If the TV signal is to be distributed to more than one receiver at several locations new considerations are introduced. If more than one station is to be received simultaneously, a rotor generally will not do because the correct position of the antenna for one station will be less than optimum, if not downright awful, for another. This is usually solved by installing several antennas on the same mast, each aimed for best reception of a particular station. (See diagram on page 15.) Such systems may use a single broadband

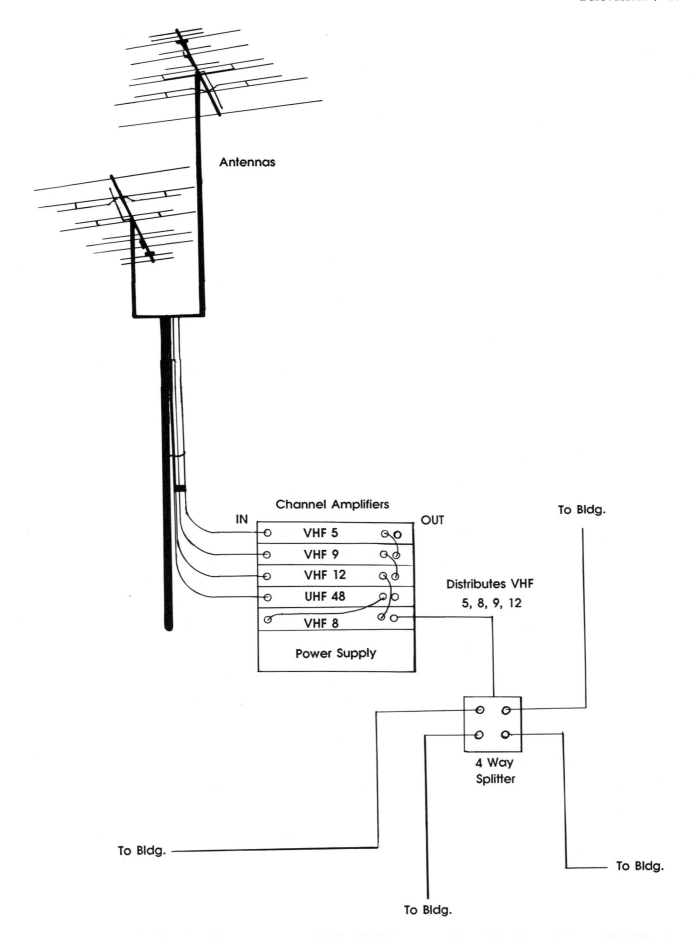

Antennas

Channel Amplifiers

IN   OUT

VHF 5

VHF 9

VHF 12

UHF 48   Distributes VHF
                          5, 8, 9, 12
VHF 8

Power Supply

To Bldg.

4 Way
Splitter

To Bldg.

To Bldg.

To Bldg.

amplifier, or there may be individual channel amplifiers for each of the channels specified at the time of installation.

Because of building code stipulations related to grounding of rooftop antennas, insurance requirements, and the danger of roof damage, actual installation of these systems is usually done on contract with a firm specializing in this kind of work. But there are some things about these systems that can be helpful to know in case of trouble.

Does anyone ever look up to see if the antenna is up or down? Very often a distribution system is inoperative and no one cares about the state of the antenna itself. But it doesn't say good things about administration or maintenance to have an antenna down on the roof, usually seen by everyone except those who would have to do something about it. Wind and roofers are the usual causes of an antenna being down. Sometimes a system quits working because the antenna is missing ... due to theft or vandalism. But it's pretty dumb to call for service, only to have to be told that the system isn't working because the antenna is missing.

Roofers or roof-climbers often break or tear out the lead-in wire. Some antenna installations are rather sloppy, with the wire trailing across the roof from whatever vertical support pipe was found to the place where the wire goes into the building. If all channel reception has deteriorated or disappeared, this is a prime suspect.

Storms with lightning, especially during summer when school is not in session and a failure of the system goes unnoticed, can cause the amplifier fuses to blow. The distribution amplifiers are usually in or near the roof, in some out-of-the-way place. In the case of broadband amplifiers, all channels would show problems. If the system uses individual channel amplifiers, only one or two may have failed, causing one or two channels to have degraded reception. (A broadband amplifier is one which boosts the signal strength of all VHF and UHF channels, or in some cases there may be three broadband amplifiers: one for low-band VHF, another for high-band VHF, and a third for UHF. Individual channel amplifiers, as their name implies, boost only one channel for distribution through the system. These are almost never in the UHF band, or if they are, they are then converted to one VHF channel for distribution through the system.)

To expand on the aforementioned converters, large building distribution systems normally convert any UHF channel to some VHF channel. This is done because UHF does not distribute well through all the cable in a large system. This usually causes great confusion among users, who will persist in tuning in the UHF number and complaining about the reception, when the reception of that channel is quite good on the conversion VHF channel.

Still another source of broadcast television signals is the satellite dish antenna with its associated amplifier, converter, and positioning and tracking system. In an institutional installation, the most versatile program selection would be to have this source coupled to one receiver in a fixed location. It would be possible to distribute the satellite reception on a conversion channel throughout a building distribution system, but all receivers would be forced to watch the single preselected program established at the controls of the satellite dish control system.

Very large distribution systems, with all VHF channels carrying some program, may show a herringbone interference pattern on some channels. This is adjacent channel interference. Often it is caused by an adjacent channel that would virtually never be used, as for example an ETV channel having the herringbone because of an adjacent channel that carries movie re-runs much of the time. If the problem is clearly put to a service person and permission given to turn down or disable the movie channel amplifier, the interference can be cleared up and the more important program distributed with minimal interference.

Upstream-Downstream is another feature sometimes incorporated in distribution systems. In such systems, in addition to the TV outlet, there may be an input jack in some select locations, or in all locations, adjacent to the antenna output jack. Output and input are rarely labeled by the installers and this results in all kinds of confusion until someone goes around and plainly labels which is which. (You find this out by carrying around a small TV receiver and plugging it in. If you get good channel reception, it's an outlet. If you get little or nothing, it's probably an input.) The installing company supplies a modulator into which must be fed the video and audio from a camera or VCR. When the modulator is properly adjusted, the program from camera and/or VCR goes up to the distribution amplifier at the "head-end" and comes back down all through the building on some specified channel. There are often some problems with the quality of the components in a system of this type, and "broadcast quality" should not be expected.

CABLE DISTRIBUTION

Depending upon the franchise agreement between a community and a cable television company, a school or institution may receive a more, or less, complicated cable distribution system within a building. Or it may just receive a "drop" with cable service to a single designated location within the building.

In the case of a single drop, the cable operator may supply a free cable channel selector unit, or it may be part of a paid service agreement like that of any householder within the cable service area. Many TV receivers and VCRs since about 1980 have included "cable-ready" tuners. Most of these can be connected directly to a community cable drop and will select many of the channels without use of a cable channel selector unit. Hedge words like "many" and "most" are necessary because there are always exceptions ... things that just don't interconnect as expected or advertised. There was also a period during which the number of cable-ready channels was increased. The number of channels selectable by a given model of TV receiver, combined with a rather wide variation in the number of channels offered by different cable systems, makes the combination a very local matter.

This might be a good time to discuss how cable systems are able to carry so many channels when receiver VHF tuners usually only offer channels 2-13. Let's start by defining what some of the letter designations stand for:

RF = Radio Frequency

For purposes of this book RF is up in the broadcast frequencies. Properly processed and given enough power, RF signals can be transmitted and received through the air. They are usually selectable with some kind of dial or channel tuning. AM and FM radio, VHF and UHF television, are all carried by Radio Frequency signals.

VHF = Very High Frequency

While it is used to designate a spectrum of frequencies used by amateur radio, TV and FM broadcasting, and other services, for the purposes of this book it is channels 2-13 television, whether broadcast or cablecast. VHF is broken up into Low Band, channels 2-6, broadcast and cable; Mid Band, designated by letters and available on cable only (the frequencies being used by FM and other services when broadcast); High Band, channels 7-13, broadcast and cable; and Super Band, being VHF frequencies above that of channel 13 and available on cable only.

UHF = Ultra High Frequency

TV broadcast channels 14-83.

Video

A broadband picture or text signal containing black to white, color and hue, and timing information necessary to produce a picture or field of text on a TV receiver or monitor capable of processing a video input signal. Video is never channel selected, although some keypad tuners do include the number combination to switch to video. Video does not include audio information. Video cannot be broadcast without first modulating an RF carrier. Video is normally interconnected by coaxial cable, and the connectors are usually different from RF connectors, at least within the same system, to reduce confusion.

Audio

A very low, relatively narrow band of frequencies that carries information for the ears.

Audio (continued)

Video and audio information are usually combined when carried on selectable RF channels. Sometimes, as in a videotape playback for a large audience, the audio output is connected to a PA system to provide more adequate sound. If video and audio are used simultaneously, a separate conductor must be provided for each. This may be a separate cord, or it may be a multi-conductor cable with a multiple pin connector on each end--often the common 8-pin connector found on video equipment.

Now back to cable distribution. Community cable systems have the same limitation regarding use of UHF as was noted earlier for building distribution systems. None of these signals is as easy to move around as house power or telephone speech, and as the frequency rises the problems increase.

The cable system is able to offer many channels because it uses at least the mid-band, in addition to the standard 12 VHF channels. Some systems also use more than one cable, easily doubling the possible selection by switching between A and B cable. Since the signal is confined to the cable and not subject to broadcast interference, it is thus possible to have a channel 4A and channel 4B.

There are a number of options for distributing cable through an institutional building. Some of these are shown in the diagrams on page 20, together with a list of advantages and disadvantages of each on page 21. This may be helpful in designing a new facility, updating an older one, or for identifying a diagram that most closely resembles an existing system as a map of potential trouble areas.

"How can I view one channel and record another at the same time off of a cable system?" is a frequently asked question. The key to the answer is in understanding where the channel is selected. In the left diagram on page 22, any channel selected on the Cable Channel Selector will be the only channel available to either VCR or TV. The diagram at right splits the cable signal to two selectors--as shown, the TV Tuner and the Cable Channel Selector for the VCR. (If the VCR tuner is "cable ready" the Cable Channel Selector could be moved to the TV leg of the split as shown in dotted lines.) The Coaxial Switch (a standard Radio Shack item usually used to switch between an antenna and cable) makes it easy to use the TV to set up the VCR or to view tapes and then to switch back to direct cable connection.

Some cable system channel selectors require the power load of the TV receiver to keep them ON and to stay on channel. If such a Cable Channel Selector is used with a VCR, it may be necessary to plug a table lamp of at least 40 watts into the selector's power outlet and keep the light turned ON to permit VCR clock-start recording of a desired channel.

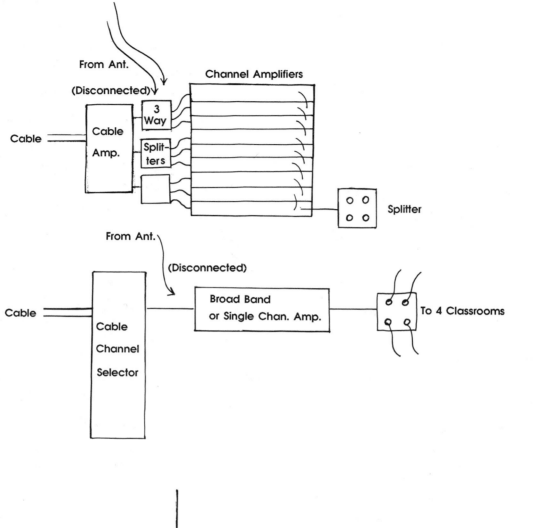

From Ant.

(Disconnected)

Channel Amplifiers

Cable

Cable Amp.

3 Way

Splitters

Splitter

From Ant.

(Disconnected)

Cable

Cable Channel Selector

Broad Band or Single Chan. Amp.

To 4 Classrooms

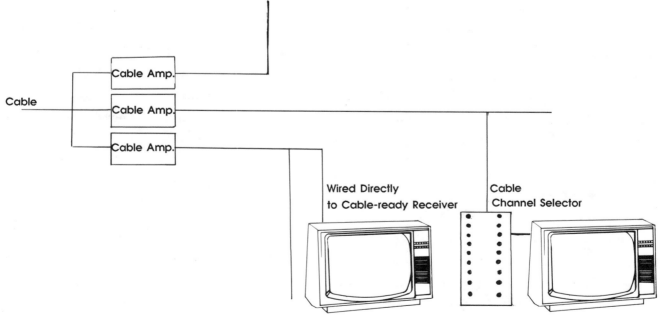

Cable

Cable Amp.

Cable Amp.

Cable Amp.

Wired Directly to Cable-ready Receiver

Cable Channel Selector

Older, 12 VHF Channel Receiver

SYSTEM	ADVANTAGES	DISADVANTAGES
Cable Interface with Existing 12 channel VHF Distribution	No change in bldg. cable Minimum conversion cost Administrative control of available channels Full use of old 12 VHF channel TV receivers	Can reduce many cable channels to 12 No renewal of an old building system Often confuses users because of channel number changes
Single Selector for Building Distribution	Total administrative control Inexpensive, especially if distribution is to just a few rooms Avoids need to provide channel selector for each TV receiver	Requires someone to make timely channel changes when needed Reduces many cable channels to 1 at any given time
Full Cable Distribution	Most versatile Makes all cable channels available to all TV's at all times Requires no control point for channel preselection Can use existing apartment house distribution equipment and techniques	Expensive to install No administrative control May require rental of multiple cable channel selectors

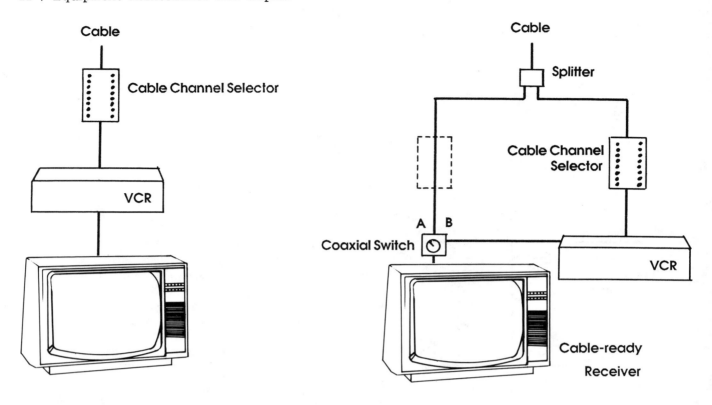

A PORTABLE DISTRIBUTION SYSTEM

It is occasionally necessary to provide video viewing for a cassette playback or computer demonstration in a location that either does not have an installed facility, or whose installed system is not adequate for the group size. While some years ago it was sufficient to provide just an audio link to handle an overflow crowd, today it is customary to handle an overflow with video and audio.

There is really no substitution for at least a 25-inch display screen in institutional applications. See the table on page 23 of diagonal screen size vs. viewing distance.

This would mean that more than one 25-inch screen should be provided if more than 15 adults seated in tablet-arm classroom chairs are expected to see the screen adequately. Furthermore, these viewing distances assume that graphics or text will have been scaled for legibility standards based on these distances. Often this is just not the case. A candidly shot videotape that includes overhead projectuals or a flip chart is at the mercy of the proportions set by those media. Often the original materials were not very readable a few rows back to begin with. Similarly, rarely is any size standard considered for computer-generated text fields. Often what is being shown is a screen from some program that, while quite adequate for one or two computer operators, is not adequate, and was never intended, for large-group viewing. Short of remaking the tape, the only solution that can offer even modest help is to provide more receivers or monitors.

After all of the above advocacy of 25-inch picture size, a set of three or four 19-inch receivers might have to be considered if mobility over a school district or campus is a factor.

Picture Diagonal	Picture Width (W)	Closest 5W	Best 7 - 10W	Furthest 14W
13"	11"	1.5	2 - 3	4
19"	16"	2	3 - 4	6
25"	21"	3	4 - 6	8

Yards or Meters

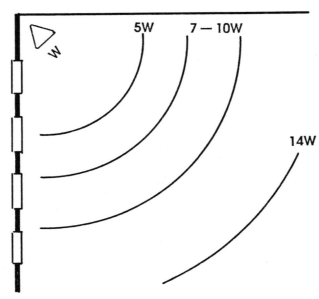

Those who have the authority to order these set-ups rarely help with the physical effort involved. Unless your organization is labor intensive--and don't overlook the liability aspect--carting 25-inch TV receivers around may just be out of the question. A sub-size picture may be better than no picture at all.

The figure on page 24 diagrams a simple distribution system. Back in the black-and-white reel-to-reel tape days, it was customary to distribute video with separate audio. This had at least the advantage of allowing a single volume control to be inserted in the distribution line near the VTR. With the advent of color, a change of connectors on the same old coaxial cables made it possible to use RF distribution (color receivers were unlikely to have a video input). A portable VCR with a strong channel 3 signal out could be used to drive the system, whether the input was video from a computer or camera, or a videocassette playback.

Audio must be controlled at each receiver, and it is usually easy to ask someone seated near each receiver to adjust the volume to a satisfactory level. If a regular VCR is being used for a tape playback, switching the input switch to "Camera" or "Line" (instead of "Tuner") will eliminate the loud rushing sound and screen snow of the tuner with no signal coming in. It is usually best to leave each volume control turned up just a little, making it possible to raise it as required without blasting everyone when the program begins.

It is also a good idea, and you should start early enough that time will permit, to huddle the receivers together and play part of the tape, or use some kind of color standard if a camera

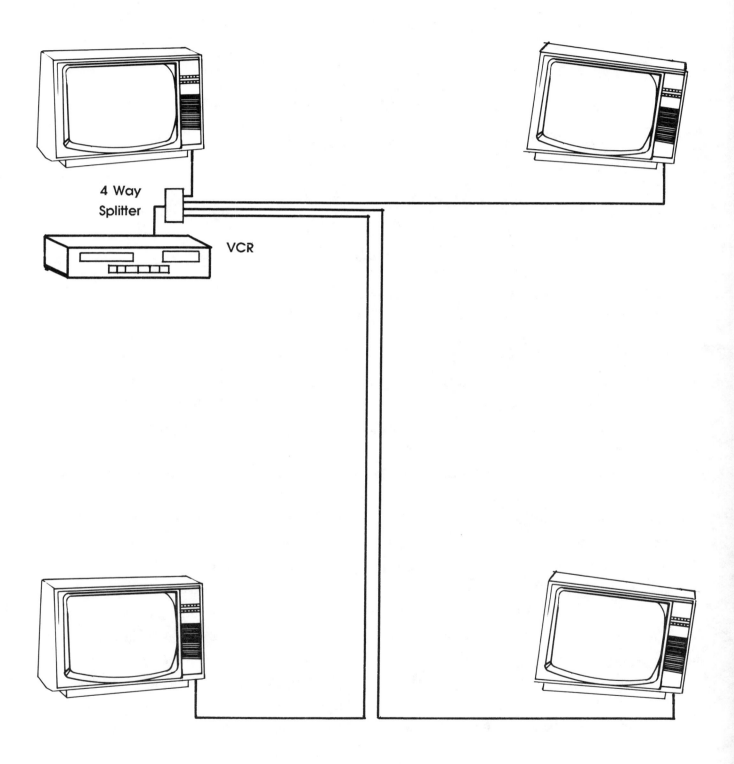

4 Way
Splitter

VCR

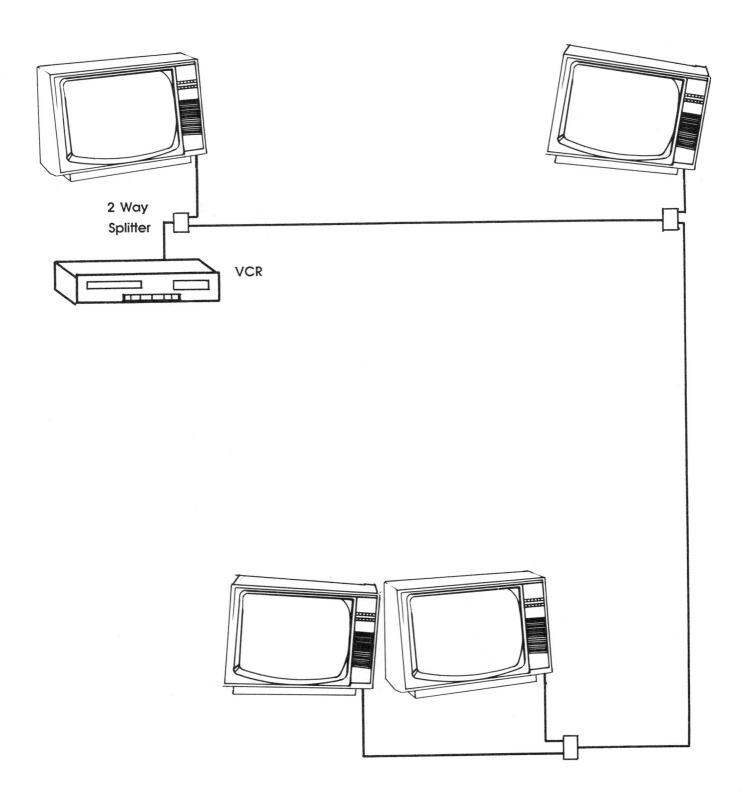

2 Way
Splitter

VCR

is involved, to adjust all receivers to be as similar as possible. It's rather disconcerting and distracting to sit in an audience and notice that all of the monitors have different hues of some given color.

Try to avoid using a series of two-way splitters as shown on page 25. Each splitter causes a slight loss of signal, and if the modulator driving the system is not very strong, the last one or two receivers in the chain will get snowy. A single 4-way splitter, even if not all the outputs are used, results in a single loss to each line.

If the VCR available doesn't have sufficient signal out to drive a distribution system, consider using a broadband amplifier. The Radio Shack type illustrated has four outputs, which

means that each receiver actually gets a boosted signal, rather than one with a splitter loss. These amplifiers can also be helpful where a single antenna is used to feed signal to several VCRs, as for multiple-time channel night-time off-air recording. With a switch the amplifier can also be a help in a situation where one channel is being watched while another is being recorded. Such a configuration would occur if an amplifier was substituted for the 2-way splitter in the diagram (see page 22) and additional VCRs were connected to the extra amplifier outputs. This would assume "cable-ready" VCRs, each with timer and channel selector set to record a different program.

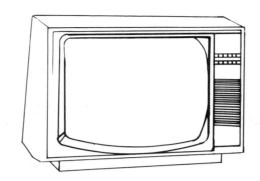

TV RECEIVERS

AND VIDEO MONITORS

The common denominator of most modern electronic media and learning/computing systems is some sort of Cathode Ray Tube display or readout. As of this writing, there is increasing use of Liquid Crystal Displays and other reflected or transmitted light devices that do not require the high voltage essential for the electrons of the cathode ray to be attracted to the phosphor screen where pictures and characters are formed. But at this time and for the foreseeable near future, CRT devices will continue to dominate the visual interface between people and machines, if for no better reason than that there are millions of them in use.

While exceptions might be found, generally a "receiver" has some form of channel-selecting tuner as part of its built-in controls. Receivers built for the education market usually have provision for a composite video input and separate audio, and include some means of switching from a channel-selected TV signal to the separate video and audio. When used in this latter mode, the receiver essentially functions as a monitor.

"Monitors" on the other hand do not have any method of receiving a television signal. Some do not include audio either. Originally "monitors" were just that--CRT devices capable of a very high quality picture, used in TV stations and studios to monitor the quality of the picture being transmitted or recorded. Picture quality is defined as "lines of resolution." In the U.S. and countries using the same system, there are 525 lines from top to bottom of a screen field. Some of these are used for synchronization and color information and are masked by the screen frame. So the maximum vertical resolution would be something less than 525 lines. Horizontal resolution is controlled by the speed with which the circuits can control the intensity of the electron beam, and in the case of color, by the matrix and color phosphors deposited on the screen. Cost rises markedly with demand for higher definition. Forty vertical columns of computer readout is fairly easy to achieve, but when this rises to eighty the need for monitor quality may become apparent. The red smear often seen during playback of a third or fourth generation copy videocassette is an example of poor horizontal resolution, although its cause is in the dubbing process, not in the receiver or monitor.

Some of the monitors marketed for use with specific computers have three inputs labeled Red, Green, and Blue, which give access directly to the color circuits for an improvement in resolution. Unless there is also a composite video input, these monitors can be considered as dedicated to the system with which they are used.

27

Aside from such limiting considerations, there is quite a lot of switching and exchanging that can be done among receivers and monitors to temporarily restore some kind of service. Most VCRs have video and audio output jacks, in addition to the customarily used channel 3 or 4 RF output. With two cords, usually with RCA to RCA plugs on the ends, you can connect a Commodore or other video and audio input monitor to the VCR and use it to watch either TV programs using the VCR tuner, or to watch cassettes. Conversely, a computer can be used with a TV receiver if both computer and receiver have video interconnection capability. Or, a VCR can take a computer video output into its "Line" or "Video In" jack and put out a channel 3 or 4 signal that can be distributed to one or more large receivers for large group showing of a computer program. As a fairly general rule these days, if the connector is "Type F" you are dealing with RF and a channel selection will need to be made. If the connectors are "RCA Type," you have a video or audio connection and a TV receiver would have to be switched to that mode. Often these switches are on the back of the receiver, near the input connectors. Some digital channel selectors have a number that must be used to connect a video input.

POSSIBLE PROBLEMS

Power

Begin with the simple question, Does the receiver or monitor have power? Sometimes it's hard to tell. Large receivers will make an audible sort of surge hum the moment they are turned on, which then quickly fades away. If you can see it, the absolute indicator of power is a glowing filament in the neck of the CRT near the back of the unit. Two other ways to get some indication of power are to advance the BRIGHTNESS control and look for some screen brightness, or to put your ear to the speaker grille, if there is one, and listen for a slight hum.

If you decide there is no power, consider the outlet first. Is anything else, (computer, VCR, another receiver or monitor) getting power? If not, it is often easier to search for a working outlet than to find circuit breakers or fuses located elsewhere in the building. If you can establish that you have power in the outlet, the next question is "Is the receiver or monitor really turned on?" Some computer monitors are particularly tricky, with switches in back, almost concealed from users. If you are convinced that the equipment is turned ON, and you still seem to have no power, the two remaining things a novice can do are look for a small red button on the back, particularly on TV receivers. This is the circuit-breaker and should be pushed in with a firm, quick push. Do not dawdle or keep pressure on the red button. If there is a real circuit problem, it is important that the button be able to snap back out. The other remaining possibility, but a very good one, is a problem with the line cord or power plug. Try bending the power plug prongs outward slightly in an effort to make them contact the metal in the outlet. Finally there is the possibility that the cord is broken, usually in the plug. The only sure way of knowing is to cut the cord and replace the plug. If a power cord is constantly wound and unwound as equipment is moved from one location to another, the cord may have broken at the receiver or monitor, or where the bending occurs. Most CRT devices have interlock power cords that assure disconnection when the back of the unit is removed. The seemingly simple replacement of these cords can become fairly high-tech by the time rivets have to be drilled out, the correct replacement cord located, and then the replacement riveted in. It may be best to try to find an exchange unit and turn the replacement job over to a professional technician.

Note: In the following sections we will repeatedly make reference to checking controls and/or cables. This seemingly self-evident source of trouble is a particular problem in the institutional setting where so many people have access to equipment. One hardly ever stays in a hotel or motel where the color of the TV is set at levels that are satisfactory. Part of this problem results from the wide variation between individuals in the perception of color, but another aspect is the psychological need to readjust anything that can be readjusted. And then there is the prankster, who cannot resist the temptation to turn or switch something, secure in the knowledge that it will give someone else fits.

No Picture

(We assume that you have proven that power is on.)

Is there any screen raster? If you cannot sense some illumination of the screen, find the BRIGHTNESS control and turn it clockwise in an effort to get some indication of screen illumination. Also, turn the CONTRAST control counterclockwise, to reduce contrast. Failure to establish any screen brightness usually suggests a high-voltage circuit malfunction ... a service job for a professional technician.

If you can get some suggestion of a screen raster, the problem is in the signal line. This too could be an internal receiver or monitor circuit failure, but more likely it is a switch error or cable failure.

Let us assume first that you are dealing with an RF signal that connects to the receiver through a Type F connector. (Monitors simply cannot handle an RF signal; you are trying an impossible match of devices.) In the case of a TV receiver, can you get reception of some TV station? Disconnect the coaxial cable and insert an uncurled paper clip into the center hole of the Type F Connector on the back of the receiver. Either keep your finger on it, or have someone else lightly hold the paper clip while you turn through the channel selector. If you don't get even a suggestion of a picture or sound, look for a switch that switches the receiver from TV to Video. Labels vary widely on these switches. One side might say TV, or Ant., or Cable and the other side might say Video, or VCR, or Video Loop. Finding and switching such a switch will probably solve the problem.

If you are trying to use an outside antenna or building antenna system, and the paper clip gives you some picture or sound, carefully look at both ends of the cable. If the fine wire in the center of the Type F connector is bent or broken, no connection will be made and no picture received. Should this be the case, remember that it is often more expedient to exchange the cable than to try to fix the one you have. If you encounter a Type F connector with the wire broken off down inside, do not despair. Find any cutting implement you can, cut the connector off the cable, cut back the outer jacket insulation and the shielding braid. Then cut away about 1/4" of the inner insulation. What you want is about 1/4" of bare center wire sticking out. Remember, this is RF, Radio Frequency. The power may not be enough to send the picture through the air, but it will stay on that center wire and it does not necessarily need a complete two-wire connection. If you want a two-wire connection, bend that

trusty paper clip to slip over the braid and the outside of the Type F connector. All of this is as dreadful as it looks, but if it makes it possible to play a tape or see a program, that's all the time it has to last. Proper repairs can be made later. Insert the inner wire into the female Type F connector and some kind of picture should appear.

If you are trying to connect an RF output device to the TV receiver, such as a VCR or computer with an established RF output, you may not have the channels matched. Most VCRs offer a channel 3 or 4 option, with switch selector on the back of the VCR. These switches are invariably recessed and some implement must be used to move them. Whether channel 3 or 4 is customarily used in an area depends upon the channels used by local TV stations and sometimes by the channel used for the output of cable converters. If a local TV station uses channel 2, VCRs in that area would probably use channel 4 to minimize adjacent channel interference. It is also common to get interference bars from two devices using the same channel, as would be the case if a cable converter output was channel 3 and a VCR also outputs on channel 3. In this case one would expect to find the cable unit on channel 3 and the VCR on channel 4. The point of all this is that you cannot assume that every piece of equipment will be switched to directly interchange with any other unit. And a TV receiver whose CHANNEL SELECTOR may be Fine Tuned for channel 3, may not have proper fine tuning for channel 4, and the FINE TUNING control (if any) would have to be adjusted to produce a picture and sound on channel 4. Neither can you assume that someone has not gotten at the back of the VCR and changed the output channel switch. Any mismatch of channel number, or improper fine tuning, can cause a no-picture situation. If TV reception can be established on any channel, it is usually just a question of calmly working through things to find out why there is no picture on the desired channel.

The composite video signal is much simpler since there is no multi-channel tuner involved. Either the video input produces a picture, or it doesn't. In the case of TV receivers with video input, if you can get a picture from a TV station using the CHANNEL SELECTOR, the probability is that the receiver has not been switched to Video. Also, watch those connectors. <u>Video</u> is virtually <u>never</u> supplied through a Type F connector. If, for some reason, you are trying to push video through a channel-selected Type F input, you will not succeed. Video cables normally have BNC, PL-259, or RCA type plugs on the ends. Video cables are coaxial cable wire, but a few feet of audio wire (with RCA plugs on both ends) would work in a pinch.

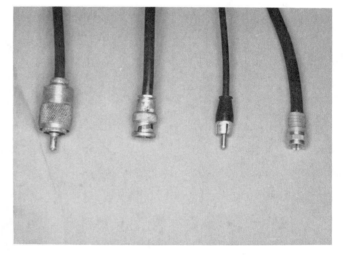

Finally, a word about inputs and outputs. You've got to be able to read. But after you read, somehow you've got to know what it means. That isn't always easy. People do seem to have an awful time keeping it straight, and getting it wrong almost guarantees either no picture or no sound, or both.

If a VCR terminal on the back is labeled VHF Ant., that's an input from a VHF antenna. It's usually a Type F connector. Its purpose is to connect an incoming signal from an antenna or building distribution system or cable converter to the VCR tuner when the VCR is turned ON. If the VCR is turned OFF, the incoming signal is passed through the VCR to the tuner of the TV receiver connected to it. The UHF antenna connection is usually a pair of binding posts to which a 300-ohm twin-lead cable can be attached. As with the VHF, there is a pass-through provision, with another pair of binding posts to connect UHF, via twin-lead, to the UHF terminal screws on a receiver. The VHF "To Rcvr.," or "To TV," etc. is the output from the VCR. No matter whether tape is playing or an incoming TV program is being monitored, when the VCR is turned ON, the channel selector of the TV receiver must be set to the same channel as the output of the VCR, either channel 3 or 4, if the program originating on tape, or selected by the VCR tuner, is to be viewed. Most VCRs will permit the signal to pass through during recording so one channel can be recorded while another is being viewed. A "VCR-TV" (etc.) switch may have to be in "TV" position for this mode. This, after all, is one reason people buy VCRs.

All of the above is somewhat different from a tendency in audio equipment to label a connector to correspond to the label on assumed associated units. Thus outputs become inputs and the inputs may be outputs and the whole thing can get terribly confused. Some people seem to get terribly mixed up if the equipment at work is not exactly like the stuff at home. Many people still persist in having someone number-tag or color-band the cables and connectors of the equipment they use. It is the authors' contention that people using this kind of fairly expensive equipment need to know why things get connected the way they do. That knowledge is only one step beyond what it takes to operate the various units, and it gives the user much greater versatility in covering the contingencies that crop up during phases of making and playing back computer programs and videotapes.

Degraded Picture

SNOW: a random screenful of white dots whizzing around like a blizzard, with more or less of the picture in the background. Snow is almost invariably the symptom of weak signal. Although it can be caused by a weak amplifier circuit within a receiver or monitor (snow is usually associated with radio frequencies and therefore unlikely to be seen on a video-connected monitor), it is more commonly the result of antenna problems, lead-in breaks or corrosion disconnects, poor fine tuning, or faulty cable and/or connectors. In building distribution systems or even in portable distribution networks, snow can result from the use of too many splitters or failure of a distribution amplifier. Splitter trouble can be identified by better pictures on receivers closer to the source, be it antenna, computer, or VCR. Use a single four-way splitter rather than a cascade of two-way splitters to reduce the problem. (See the earlier discussion of distribution systems.) In building distribution systems of the separate channel amplifier type, snow on one channel but not on another is an almost sure indication that either the snowy channel amplifier has a problem, or the associated antenna ia aimed the wrong way, half-or-wholly disconnected or stolen.

GHOSTS: those lightly outlined images that repeat the main image, to the right of that main image. They are almost always an indication of a reflected signal, arriving just after the primary signal because the reflected signal path is slightly longer than the direct primary signal path. Turning the antenna is the customary way to reduce, if not eliminate, ghosts. While ghosts are usually associated with RF transmissions, they can also occur with video distribution systems. In this case they result from a reflected signal coming back down the coaxial cable from the last receiver. The cure for this problem is to add a 72-ohm terminating resistor across the cable at the last receiver. Since the connectors and cable are totally shielded, it is necessary to put the resistor in a small box with appropriate input and output connectors for the video system in use. Adding a small switch lets you put the resistor in or take it out ... whichever produces the clearer picture.

In urban settings where signal strength can be very great, there can be other causes of ghosts. One local hospital is located almost at the base of a TV antenna tower. That station's signal is so strong in that area that the transmitted signal permeates the building, combining with the internal cable-distributed signal to produce a heavily ghosted image.

As with the video distribution discussed above, ghosts can be caused by pranks and vandalism within a building distribution system. The TV outlets of distribution systems are terminated with 72-ohm resistors. If an outlet is damaged or shorted, some form of picture degradation, at least along that branch of the system's cable, is likely. This would most likely appear as ghosting, or as a weak signal resulting in a lot of snow.

BARS, HERRINGBONE PATTERNS, PULLING, AND FLAGGING: these annoyances degrade an otherwise good picture. Sometimes they result from problems within the receiver or monitor, but more frequently bars and herringbone patterns are symptoms of some kind of interference. Some computers generate spurious signals that can produce patterns on a nearby TV receiver. Motors starting and stopping on the same general power line may cause streaks of one sort or another, which come and go depending upon whether the motor is running or shuts off. Adjacent channel interference in the form of bars or herringbones is fairly common in large distribution systems and may not be able to be eliminated. If you have the option of video or RF connection, changing from RF (which is always a little easier because it doesn't require the separate audio) to video plus audio will often eliminate these interference problems. Pulling and flagging, in which part of the picture becomes elastic and stretches left or right, is usually some kind of frequency problem. If the receiver or monitor has both HORIZONTAL and VERTICAL HOLD controls, these must be set as near "center" (the point midway between those control settings where the picture rolls or pulls one way and then the other) as possible, allowing some latitude for changes in the incoming signal. Slow and delicate adjustment of TRACKING and/or SKEW controls on a VCR will often effect at least a partial remedy or minimize the flagging. Again, the use of separate video and audio connections, if possible, will often save the day when the RF connection just seems unable to settle down.

COLOR: fading, poor hues, over-saturated colors are primarily receiver problems, although the first is not. Like snow, which color fading often accompanies, the tendency to almost lose color during TV reception is a sign of lack of signal. Sometimes a passing aircraft deflects the radiated RF television signal and the color fades, accompanied by multiple ghosting and even loss of sync, resulting in a quick roll or pull, then a sudden return to normal picture. Color problems are often particularly acute when rabbit-ear antennas are used within an institutional building. The only real solution is to get more signal to the receiver, either from a roof-top antenna or whole building distribution system. Even then, if the antenna is in some compromise position to receive a signal from two stations, atmospheric conditions (clear or rainy days make the picture better or worse) and aircraft can interrupt the color. There is also no absolute assurance that the station is always up to full power. If it is a question of balancing color hues and setting degrees of saturation, the usual advice is to turn the COLOR control counterclockwise until you have a black and white picture. Then adjust the BRIGHT-NESS and CONTRAST controls for what you consider to be a good black and white picture. When you are satisfied that you have a good black and white picture, slowly advance the COLOR control clockwise until the picture barely tints, then advance it just a little more. Having established all these settings, you can work with the TINT and PICTURE (or whatever the color saturation controls are labeled) controls to get the best possible flesh tones, blue sky and green grass. Slight readjustment of the BRIGHTNESS control may also help at this point. Most people grossly overdrive these circuits, producing saturated colors that are unnatural and tend to fuzz up the whole picture. Excessive brightness that cannot be toned down and results in a rather washed-out looking picture is likely to be a picture tube problem again. TV service shops can often make a rejuvenation treatment that will gain some more time, but replacement will ultimately be necessary.

For a really definitive check on VCR/Receiver/Monitor adjustment and performance, consider buying the Society of Motion Picture and Television Engineers (SMPTE) "Video Tape Cassette for Receiver/Monitor Setup" V2-RMS in the cassette format of your equipment. No more expensive than many pre-recorded tapes, the color bars have been recorded and are certified to be within established standards. In addition to its known quality, the tape is arranged to facilitate a quick check for overall performance of the system.

The SMPTE Receiver/Monitor Setup cassette would be especially useful if your installation includes more than three or four VCR/Receiver/Monitor units. It would also provide a reference at the time of a purchase decision.

Audio

The audio system seems anticlimactic after all the picture problems, but is an essential component of most video and many computer programs. If there is no audio, put your ear near the speaker and try to decide whether or not there is a little hum. Also, with your ear away from the speaker, try advancing the volume control and again getting near the speaker in an effort to hear some hum. If you hear absolutely nothing, look for a headphone jack, and if there is one, try to get sound in a headphone. Since headphone jacks also often disconnect the speaker, just plugging in and removing the headphone plug may restore the audio from the speaker.

If you get sound in a headset but nothing from the speaker, it is probable that the speaker is disconnected or burned out. Checking this, even just to look, would require opening the cabinet of the receiver or monitor. For safety reasons because of the high voltage inside these units, we must insist that this is work for a professional technician.

If you can hear a little humming from the speaker, try to find the AUDIO INPUT connector. With the volume control turned at least half-way up, insert an uncurled paper clip no more than a half-inch into the connector and listen for a fairly loud hum. That's you humming, though not your usual tune, but rather the 60-cycle hum your body picks up from all the electrical wiring in the building. We use it a lot for testing. If you get a hum there, plug the audio cord back in (assuming separate audio and video connection) and pull out the other end and touch the center pin with your finger and again try to get things to hum. (Note: If you are dealing with a receiver, you will have to switch to the VCR or Video inputs to have these work.) If you get hum through the audio cable, you must suspect the source device. Failure to get hum through the cable would indicate an open or shorted cable.

Distorted sound, especially in television reception, is almost assuredly the result of poor fine tuning of the channel selector. Often this tuning is a compromise of best sound considering the picture, or best picture, considering the sound. Again, a good signal helps a lot. Weak reception from poor or inadequate antennas messes up everything. When separate audio and video connections are used, there is almost no possible cause of distorted sound other than a poor source (bad videocassette audio; poor computer output) or cranking up the volume on the receiver or monitor beyond what the amplifier and speaker can handle.

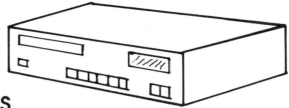

VIDEOCASSETTE RECORDERS

A properly functioning VHS or Betamax videocassette recorder should play back a tape made off broadcast television or camera with a quality virtually indistinguishable from the original program during recording. Much of the incredible public acceptance of these machines can undoubtedly be attributed to their consistent reliability and ability to perform to the above standard. Not only has their original out-of-the-box performance been amazing, but so has their ability to continue that level of performance for several years before servicing is required.

VCRs contain precision mechanisms and quite complex electronic circuitry. Current models represent several generations of evolutionary engineering development and refinement. Given reasonable care they will return extraordinary service, all things considered.

About the most maintenance that can be expected in the average institutional setting is preventive maintenance, and that's not all bad. Our users are too often a careless lot, expecting performance upon demand but having little mechanical sensitivity and unwilling to do those little things that would assist in the assurance of reliability. In other words, with the VCR below the chalk ledge, they will erase the board while the VCR is not only uncovered, but with the cassette door open, allowing all that abrasive cloud of chalk dust to settle into the mechanism. Trying to alter such placement problems, plus some periodic cleaning, can prolong good service considerably.

Bolder technicians might try installing a belt replacement kit. The number of belts, as well as accessibility, varies from model to model and almost always requires some dismantling of the mechanism. But, given proper tools, it can be done successfully.

The most extreme VCR field effort would be a video head replacement. Because these heads are now supplied as a complete pre-aligned drum assembly, it is possible to change them without elaborate alignment jigs and extensive test equipment. You should know, however, that head replacement can necessitate some circuit compensation, which would definitely fall in the province of a professional technician.

There are no symptoms that a VCR needs preventive maintenance. It's the prevention of symptoms that preventive maintenance is all about. Preventive maintenance is usually performed by the calendar or by the clock, as in during summer vacations and Christmas breaks, and/or after 200 or 500 hours of use. It might also be wise to factor in the environmental conditions under which the VCR is expected to operate, as in the aforementioned chalk-dusty area, or a machine used for training in an area heavy with chemical dust or where the air is full of oil particles from rows of operating machine tools. High ambient heat will cause a VCR to get quite hot while operating, and heat hastens the failure of the neoprene belts. It is possible to conceive of situations so bad that it might be necessary to have six or seven machines for five training stations, making it possible to circulate units for preventive maintenance constantly.

A good preventive maintenance routine should consist of three parts: Cleaning, Visual Check of Mechnical Components (particularly belts and idler wheel drive surfaces), and a final Normal Operating Check.

Before going into maintenance there are a few operating hints that might be noted:

1. VCRs are compact units, normally metal encased. There will be almost no leakage of a TV signal into the unit, so an antenna is absolutely required for the channel selector to work. This is contrary to TV receivers, where a composition or plastic back, even if the case is metal, will often allow enough signal into the tuner to permit at least some suggestion of a picture.

2. The tuner of a VCR may have greater or less sensitivity than the tuner of an associated TV receiver. Do not expect picture quality to be identical when either of the tuners is used. If either the VCR or TV has fine tuning capability, and the picture is better on one than the other, some fine tuning might be needed.

3. In the situation discussed in 2 above, don't overlook the need for accurate fine tuning of a TV receiver to the channel 3 or 4 output of the VCR. For example, let's say channel 5 looks good when selected on the TV receiver, but less good when the channel 5 button is pushed on the VCR. An attempt to fine tune channel 5 on the VCR does not materially improve the picture and/or sound. Remember that the VCR is converting channel 5 to channels 3 or 4 for monitoring on the TV receiver, and therefore the frequencies of the 3 or 4 output of the VCR must be accurately aligned via FINE TUNING into the TV receiver.

4. Many VCRs have a TV-VCR or TV-TAPE switch that must be in the VCR or TAPE position to play a tape. Especially during an initial set-up demonstration, as one works through the possibilities, this switch is left in the TV position and it is impossible to play a tape. Panic time! Just take a moment to double check all those switches and selectors. Similarly, if the demonstrator has been working back and forth between the TV tuner and the VCR tuner, it is often impossible to see the tape playback because the TV channel selector was not returned to channels 3 or 4. And which is it? In an institution where channel 3 is normally used as the output channel for most VCRs, it can be very confusing to suddenly encounter one where, because of some associated equipment interference, channel 4 is being used between VCR and TV receiver/monitor.

5. When several TV receivers are connected to one VCR through a splitter for large-group showings of tape, there will be snow on the receivers and a loud hiss from the speakers when the tape is not being played. This is open channel noise coming from the tuner. If the VCR has a CAMERA-TUNER or VIDEO-TUNER or LINE-TUNER switch, move it to the Camera, or Video, or Line position, which will usually blank the screens and hush the hiss. Since there is no common audio volume adjusting control for the whole system, leave each volume control just slightly advanced and ask someone near each receiver to adjust the volume control for a comfortable level in that area when the tape playback begins.

6. If a camera is connected to a VCR, don't overlook the start-stop trigger or button on the camera, which may be over-riding the normal tape start-stop on the unit itself. Portable VCRs can be especially confusing on this point. Pressing PLAY or RECORD/ PLAY and seeing the tape load up out of the cassette and then stop is almost a sure indication that the machine is in STOP or PAUSE from the camera control.

CLEANING THE VCR

Begin by getting the things you will need for the cleaning job:

*   The necessary screwdrivers for removing the covers of the machine
*   A source of compressed air
*   Some long-handled cotton swabs
*   A few facial tissues
*   A can of video head cleaner and some plastic foam or chamois swabs. (Use only cleaner plainly labeled for video heads.)
*   A bottle of isopropyl alcohol
*   Some rubber revitalizer, if available

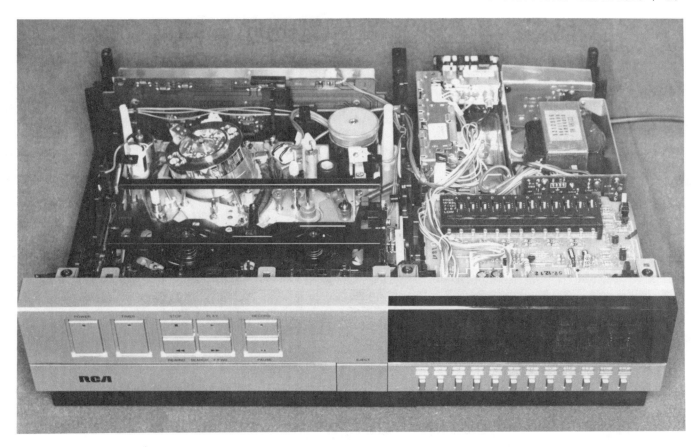

Remove the screws that attach the top cover of the VCR. These are usually plainly visible, but if removal of the obvious screws does not release the top cover, it may be necessary to remove the back cover panel to gain access to any additional screws. Cover design also often requires sliding the cover in some direction, such as back, to disengage it from the front panel.

Turn the machine upside down and remove the screws attaching the bottom plate. Always watch for any screws that are longer than others and be sure they get back into the holes from which they came. Remove the bottom plate.

If the VCR is a top-loader, remove the
rising cover of the cassette compartment.
Again it may be necessary to slide the
cover forward or back to disengage it after the screws are removed. There will
also most probably be an internal metal
shield over the video head. Some of
these are snap fitted, others use screws.

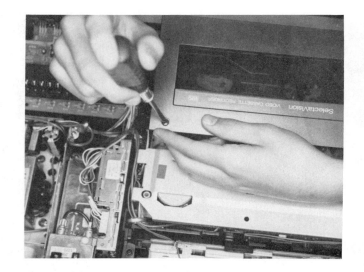

At this point you should have a fully visible mechanism. A VHS (top) and a BETAMAX (bottom)
machine are shown on page 41. Note the call-out of specific components for future reference and
try to locate them in the machine you are cleaning.

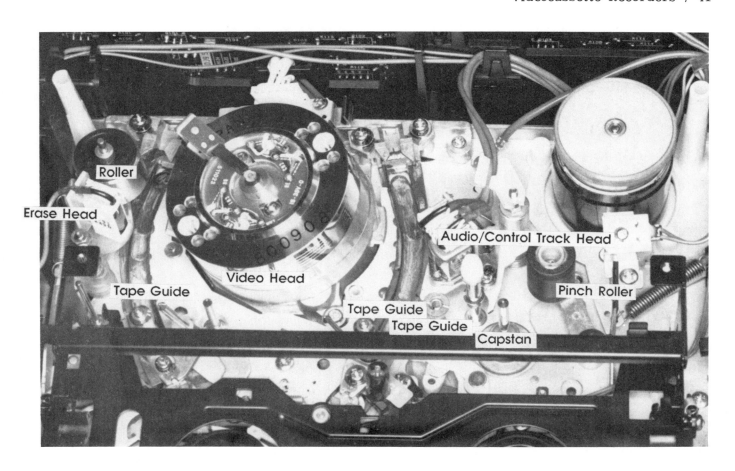

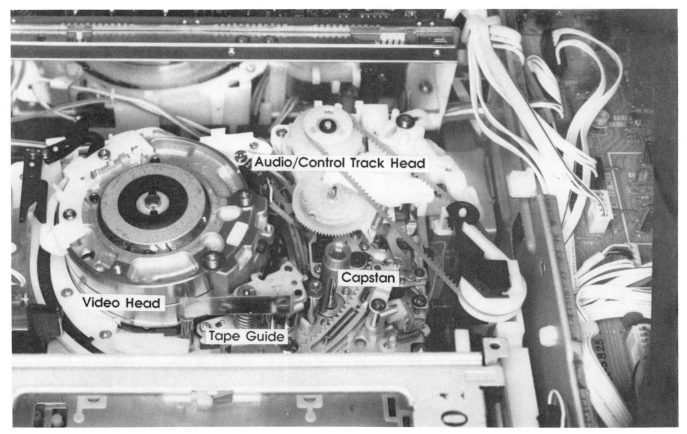

Moisten a facial tissue and/or a cotton swab with alcohol and carefully wipe out any accumulations of dust and lint that you may see, particularly at the lowest levels of the mechanism that you can conveniently reach.

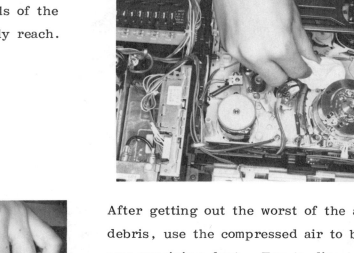

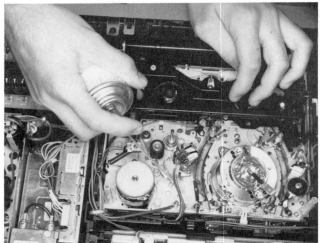

After getting out the worst of the accumulated debris, use the compressed air to blow out any remaining dust. Try to direct the air such that it will blow stuff out of the mechanism. This is also a good time to wipe out the bottom plate.

Next, turn your attention to the pulleys, drive wheels, and other parts along the belt paths. Using a cotton swab moistened with alcohol, carefully wipe off these parts.

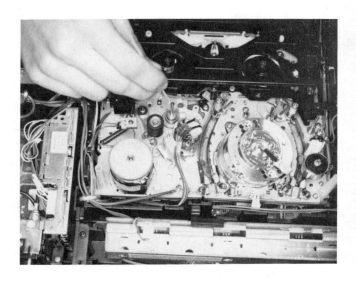

Do not get the parts wet with alcohol--we do not want to alter the lubricants nor do we want alcohol all over the belts.

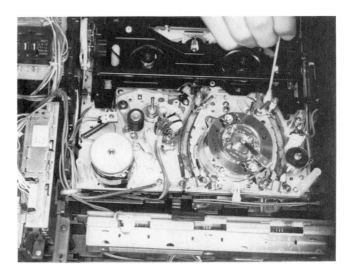

If you have some rubber revitalizer, use it now on the surfaces of belts, pinch rollers, and idler wheels. As usual, don't use too much solution. Do apply more pressure than just a gentle wiping ... You want to cut through any hard surface that may have developed as a result of oil vapor, heat, and age. A well-cleaned rubber surface loses its hard shiny appearance and looks dull grayish-black. If you do not have rubber revitalizer, you will have to do the best you can with video head cleaner.

Finally, and with great care, moisten one of the special chamois or plastic-foam swabs and carefully clean the video heads--the small black protrusions in the slot of the bright metal head cylinder. Always rub in the direction of the slot, never up and down. There may be from two to five of these heads, depending upon the features of the particular machine. The heads can be rotated by hand to give better access to each one.

When you have finished cleaning the heads, remoisten the chamois or plastic foam swab and clean the stationary erase and audio/control heads, and last of all, the tape guide posts. These posts have a tendency to cake up with fairly heavy oxide deposits from the tape, and all of this should be removed. Resist any inclination to use a metal implement to scrape them, as this could scratch the guide and ultimately remove even more oxide from the tape. Soft wood such as matchsticks and toothpicks would be ok, but the best bet is to just keep rubbing with the special swab, adding cleaning fluid from time to time. Do try to get every trace of oxide off of these guides.

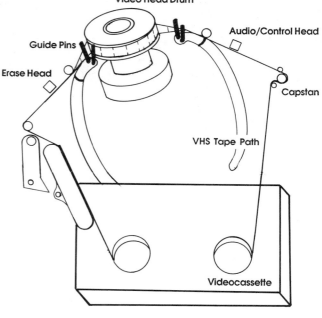

Internal cleaning is now complete. Replace the bottom plate, plug in the VCR, insert a cassette, and try the various machine functions. (Do not try this in an area brightly lit with daylight or close bench lights. Several tape transport functions are under photo-electric control and high ambient light when the cover is off the machine can confuse this control system.)

There are three machine functions to observe particularly. Looking through the cassette window, does Fast Forward appear normal as to speed and constancy of rotation? Can you start and stop the function when the right spool is empty and when it is almost full? Failure of any of these tests could indicate a need for belt cleaning or replacement, adjustment of the spindle clutch (a professional service job) or, in some designs, a slipping or hardened rubber idler or pinch wheel.

Again looking through the cassette window, does Rewind appear normal as to speed and constancy of rotation? Can you start and stop Rewind when the left spool is empty and when it is almost full? Failure of any of these functions would indicate the same kinds of problems as a failure in Fast Forward.

With the machine running in Play mode, look at the tape path through the guides and over the heads. In a properly operating VCR with a good cassette inserted, the movement of the tape is so smooth as to be barely perceptible. If the tape twists back and forth from top to bottom, or if there is fluttering of the tape along its path, stop the VCR and try at least one other cassette to determine whether there is a machine problem or if it's the tape. This is also a good time to double check all the tape guides and contact points to be sure you have cleaned all of them. If you have indeed cleaned all the guides well, and all the tapes you try exhibit the same problem along the tape path, there is probably a stretched belt to one or both spindles, or the back-tension needs adjustment--again a professional service job. Above all, do not return a VCR with this kind of problem to a use area unless you are prepared to deal with serious tape destruction problems.

If the tape flow looks good, and functions seem normal, turn off the VCR, replace the video head cover, and then replace the top of the machine. This cleaning and check should assure another six months to one year of normal operation.

REPLACING VCR BELTS

Before including this procedure in the book, we asked two above-average high school students, both good with hand tools and one with some audio cassette recorder repair experience, to try both belt and head exchanges, without reference to a service manual. The VCR was an RCA Model 255 VHS machine.

Complete kits of belts for VCRs are packaged and sold by several companies, as well as through the parts centers maintained by the manufacturers. Matching the new belt with its corresponding old one should be done carefully. Old belts will have stretched from age and service, and in this kit two of the belts were very close to being the same diameter.

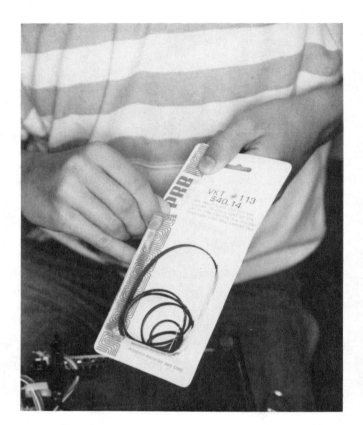

Getting access can be at least as difficult as replacing the belts. It is often necessary to remove the front panel to release a circuit board covering part of the mechanism.

Remove any screws that may be holding the circuit board down. In this model (page 47, left), a circuit board completely covers the bottom of the mechanism. Other models may appear almost completely accessible (machine on right, page 47), but still require lifting the circuit board to get access to screws that retain parts covering the belt path, as at the bottom center of the unit shown right, next page.

If a circuit board does not hinge or lift up freely, continue to look for screws you may have missed within the board area. There may also be a few wires in the way, and/or retaining clips around the sides. Never force a board because the fine etched traces are easily

cracked. Similarly, use a tool that fits the screws and make every effort to prevent the tool from slipping and gouging the board surface. It's not as difficult as it sounds, but does require care.

American Phillips screwdrivers, that come to a point, are not quite right for the ISO cross-slot screws common to VCRs. Grinding or filing the tip enough to remove the point will give a better fit and reduce danger of slipping or tearing up the screw slots.

Concentrate on one area at a time, matching and replacing belts as you go. Further loosening and disassembly may be necessary to get old belts out and new ones in, and assemblies should be resecured as the work is completed.

While belts are off the pulleys is a good time to clean the pulley grooves. Use a tissue or Q-Tip moistened with alcohol or head cleaner. It is better to work around the pulley groove than to turn the pulley, since some of the pulleys are connected to parts of the mechanism that are best not moved by hand.

Some shafts have more than one pulley, requiring sequential installation of belts in the right order. Do this carefully, especially when belts are close to the same size.

Moving to another area, remove any bridging parts of the mechanism to give free access to pulleys.

Hold pulley to prevent turning and clean the groove or flat surface that makes contact with the belt. Allow cleaner time to dry before installing new belts.

Snap new belts into place. Avoid unnecessary stretching. If it is necessary to rotate a pulley to get the belt in the groove, rotate it back to the starting position by hand.

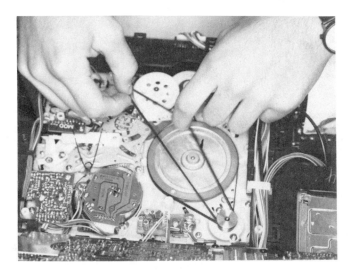

Some belts and pulleys move parts of the mechanism that are automatically controlled. It is best to start the machine as near the original position as possible to avoid confusing the control system.

Turn the machine right-side-up and replace any belts on the upper side of the mechanism. These will usually include the belts that drive the spindle turntables that wind the tape in the cassette. Access is cramped and it may be necessary to use an implement, like the wooden stick on a cotton swab, as a leverage tool to get the belt into the pulley groove.

Belts are not the only rubber drive elements. You can see the large black rubber pinch roller near the capstan on the left center of the photograph above. Rubber revitalizer can be used on this roller a number of times, but eventually the roller will have to be replaced. Pinch rollers are not included in belt kits and must be ordered separately.

Similarly, some designs make use of rubber drive tires around various wheels. These rely upon friction against a nearby drive wheel for operation. Revitalizer and cleaning of the adjacent metal drive wheel will bring them around for several times, but eventually they will need replacement.

A reasonable policy would be to try one belt replacement within the institution, and to follow it the next time with a service shop overhaul. Alternating this way would combine economy with periodic professional service attention.

HEAD REPLACEMENT

Replacing the head in a VCR is a gamble at best. It is sometimes possible without other test or alignment equipment in VHS machines. It is virtually impossible in Betamax equipment without additional alignment tools. If the institution has five or six of the same model machine, the gamble is reduced because failure to fix one machine with the replacement head would still offer several chances that one or more of the others would be less critical and that the head could be successfully installed in another unit.

Although we encountered no difficulty with our trial procedure, and the machine worked as well after head replacement as before, we did have a tough time removing one screw that was really incidental to the effort--the screw that secures the contact arm that touches the center of the drum shaft.

There is some variation in the head connecting technique. An extra pair of hands is helpful, with one person holding the head drum from rotating, while the other unsolders the

four connecting wires, if this is required.
Do not use a soldering gun when working
this close to any kind of tape head because
of the strong magnetic field generated by
the soldering gun tip. Cotton gloves are
recommended when handling the polished
drum surface. If the old heads are really
bad, reserve extreme caution for handling
the new replacement head.

After unsoldering the head wires
(if any), very carefully remove the two
screws in the top of the drum. You may
want to touch-solder the fine wires together to get them through the holes in
the new head drum easier.

The replacement head comes nicely
boxed to prevent damage. The side of
the box also contained a number of
handling and replacement suggestions,
but the translation to English left a little
to be desired.

Because the whole head drum assembly is mounted at an angle, it may be a
little difficult to get the replacement head

drum at the same angle, but a little gentle rocking will drop it into place. Do not use excessive
downward force on the screwdriver either when tightening or loosening the head drum screws.
This assembly is well mounted in the cast frame of the machine, but alignment is critical and
too much force could cause misalignment.

The new head is shown in the box above the disassembled head drum. While heads may be common to several brands and models of VCR, the replacement head must be one specifically for the machine in which it is to be installed.

Use cotton gloves or a facial tissue when handling the replacement head, both removing it from the box and placing it into position in the VCR.

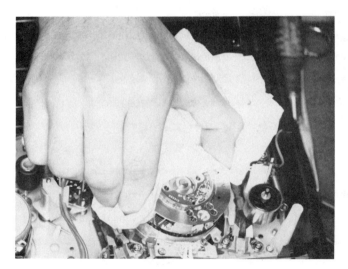

When the replacement head drum is in place, reinsert the two screws and tighten. Then solder the wires back into place, paying attention to the color code printed on the circuit board. Try to make your solder joints look as much like the original factory job as possible ... nice round little drops of solder. Remember, this assembly revolved at high speed. Wires should be dressed down against the board as they were originally, and solder drops should be as equal as possible so as not to unbalance the spinning drum.

Finally, use a head-cleaning swab and some head cleaning fluid to thoroughly clean the polished surface of the drum and remove any possible finger grease that may have gotten on it.

Restore as much of the machine as is required to play a tape. Insert a pre-recorded tape, and if it plays well, you lucked out and successfully replaced a video head.

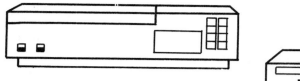
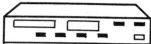

VIDEODISC AND COMPACT DISC

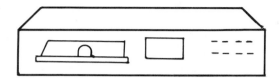

DIGITAL DISC PLAYERS

Though it is an emerging technology with a number of changes envisioned, and probably not yet in very wide institutional use, we felt this book would not be current without some discussion of digital disc systems. The technology is quite complex although the concept is fairly simple and a logical evolution using aspects of several earlier recording systems. The digital factor is new in recording and reproduction, if older in the area of radio-teletype communication. Appendix V offers an overview of details of how digital recording compares with some other systems.

This discussion stresses Compact Disc (CD) audio machines because their wide acceptance is developing some literature and experience with what goes wrong. Videodiscs are similar, if not identical, and comparable problems and remedies can be inferred. Our concern here is to be sure that you not call for professional service help when none is needed ... the usual operator level type problems. If you have a real internal problem these units will require a trip to a service shop. Inside the cover of most of these machines is a notice similar to this: "DANGER: Invisible laser radiation when open and interlock failed or defeated. AVOID DIRECT EXPOSURE TO BEAM." That sort of thing should be warning enough.

DISC RELATED PROBLEMS

These are a very common source of trouble. For a while we can expect a lot of putting the CD in upside down. And there is a good logic behind that error, in that phono discs are played from the top, but CDs are read from the bottom. A person seeing the label printed all over the face of the disc just knows that side can't play, so they put that side down. Immediately the player is lost because it can't find even the most basic opening indexing information.

Poor handling technique can leave the disc smeared or scratched, and these interruptions of the laser beam optical reading system can cause skips, tracking problems, or stopping the machine because it cannot decipher the information it is getting.

Much has been written about the durability of discs and how, since the outer surface does not contain the information, the actual recorded surface is protected. There is also the matter of beam focus, which is beyond the outer surface and tends to ignore slight scratches. All of this is true. But it is also true that some people have been gross beyond belief in the way they put drinking glasses down on phonodiscs, leaving rings. Many people persist in handling film, phonodiscs, etc. by the flat surfaces rather than the edge, leaving whatever sticky mess is on their fingers all over the medium. Digital disc recording is not going to tolerate these sloppy practices. If your facility circulates digital disc media, the discs need a visual examination of the playing surface(s) after each use. (See the part of Section III: Materials Maintenance and Repair for cleaning technique.)

Around 1987 there was a high rate of return of new CDs. The disc-making industry lacked production capacity to meet demand and was rushing discs to market with poor quality control. When you read through the complexity of the manufacturing process, you marvel that any discs get to market at all. So, if you have trouble with a new disc, it might be wise to suspect the disc before jumping to conclusions about machine failure.

Phonodisc plays from top side

CD plays from bottom side

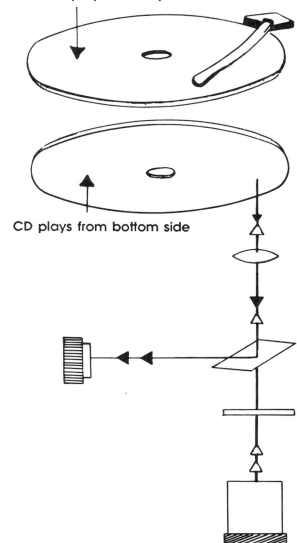

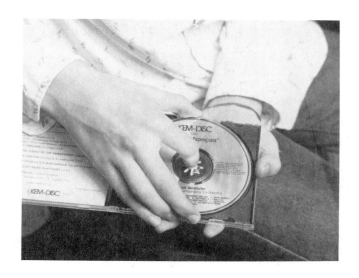

There is some difference in the way various makes and models of machines achieve the correction signal responses that keep speed and tracking where they are supposed to be. It is therefore conceivable that a disc that doesn't work well on one brand or model of machine might work better on a different brand or model. But given properly operating machines and a good disc, it should play on any machine.

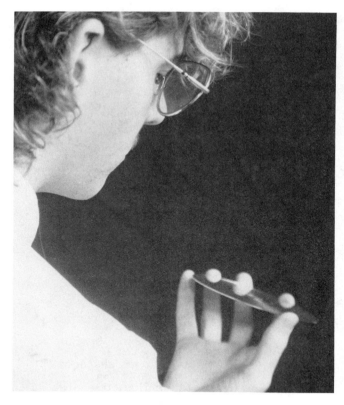

DISTORTED SOUND

Sound distortion can result from a number of causes. First, visually check the playing side of the disc, this time looking for an area that appears dull on the surface. Cleaning abrasion or a small smear of something could cause distortion.

Do not suspect a bad laser, even though this could be the problem. The industry expects the lasers to last at least 5,000 hours, about six years of normal home service, so there should not be a counterpart of the dirty needle or worn needle common to distorted sound from phonodiscs.

There can be a problem with dirt or condensation on the objective lens just below the disc. If the machine is brought in from the cold into a warm, humid room, it is even possible to have condensation throughout the optical system. A blast of compressed air aimed at the objective lens--if you can get to it--is the solution to the first problem. Overall condensation requires overall machine warming. The recommended procedure is to plug the unit in and turn it ON for about an hour, during which the heat of the electronics will bring the system up to room temperature and usually evaporate any condensation.

Another source of distorted sound is connection to the wrong input jack of an associated amplifier. CD players, like most amplified accessories, put out a fairly high level signal. They therefore require plugging into an Auxiliary or Tuner input. It would seem reasonable, since the CD is a form of phonodisc, to plug the player into the Phono input. But the gain at this input would be far too high, overloading the circuits and causing distortion.

Finally, there is the potential problem of just overloading an amplifier and/or speakers with the very high dynamic range of CD sound. Unlimited by constraints imposed by a moving needle as was the phonodisc, the CD reproduces (some contend exaggerates) the full dynamic range of the original sound. If the volume level is set for comfortable listening to a very soft passage, the power of a loud section may just overload everything, resulting in distortion if you're lucky, and blown out components as the worst case condition.

PLAYER DOES NOT OPERATE AT ALL

This problem raises the usual question, Have you got power on the thing? Most CD players light up like Christmas when they are first turned ON. If nothing happens, suspect a dead power outlet, poor power plug connection, or broken cord or plug.

Next, check to be sure the disc is inserted label side up. An upside-down disc will usually keep the machine from operating at all, although it should show indications of power ON.

If the player is new, or has just been moved, turn everything OFF again and tilt the machine back so you can see if the transport screw(s) have been removed from under the disc player area. Before moving a digital disc unit (as in a car or moving van) the unit should be operated to STOP, the disc removed, and the drawer closed. This will assure that the laser/ reader assembly moves to its rest position. At this time the transport screw(s) should be installed to prevent accidental movement of the laser/reader assembly during moving. But these screws, or screw, must be removed for the unit to work. They are usually clearly differentiated from assembly screws by labeling or by being of some obvious color like red or green.

If there is power but no sound, don't overlook the associated amplifier and speakers. Are the speakers properly connected? Is the input selector switch set to correspond to the input to which the CD player is connected? And is the volume control turned up at least a little?

INDEXING PROBLEMS

These can result from simply giving directions to the player faster than it can process that information, or before it is ready to. Older units can be rather slow responding to the push-buttons on the front. It takes a moment after a disc is inserted for the player to read the index and get ready to accept selection information. Just cooling it a little and taking the time the player needs will reduce or eliminate selection errors.

Indexing problems can also result from a speck of dust or lint in a crucial area of the disc playing surface. Having repeated problems with the same indexing point could be indicative of a disc-related fault.

INTEGRATED CIRCUIT RELATED TROUBLE

Scanning some of the service literature, most of the actual troubles within digital disc units seem to be symptomatic of failure of an integrated circuit. Even if you were equipped and skillful enough to dig out and replace one of these ICs, it is doubtful that you could find a replacement on the market because these are special ICs designed and developed for the special needs of digital disc technology. Many are probably proprietary to a specific brand because they make it possible to offer the features expected to give a competitive advantage. Returning the unit to an authorized service technician is not only the easiest solution to these problems, but probably the wisest in the long run, regardless of the size of the repair bill.

From a preventive maintenance standpoint, there is no substitute for a clean environment for digital disc technology, be it video or audio. Chalk dust and these machines is an impossible mix, though they might be considered an essential mix of some teaching/learning environments. If the setting is modern enough to include high-tech, it is modern enough to incorporate a large writing surface that uses the newer liquid markers that can be erased without leaving dust blowing around the room.

Nor should the cleaning of high-tech equipment be left to normal custodial services. While it may be necessary to direct custodial procedures to minimize stirring up dirt when cleaning floors and furniture in rooms where equipment is located, the actual cleaning of the equipment should be carried out, frequently and on schedule, by personnel with the sensitivity to know what they are doing, and the implications of doing it wrong.

Just simple little things like periodically wiping out the bottom of the CD loading drawer and keeping the drawer closed when the machine is not in use can result in marked improvement in reliability. It's a small price to pay for the superior digital disc performance.

COMPUTERS

CENTRAL PROCESSOR

Of all the elements of your computer system, the Central Processor should give you the least problems. You will find far more of your time devoted to malfunctioning disk drives, printers, monitor, connectors and cables, and particularly software, than you will the Central Processor. Problems characteristic of the Central Processor can often be prevented with a little prior "tender loving care." These problems are usually caused by outside factors that contribute to the total failure, or at least malfunction, of the electronics.

Keeping this component of your computer clean is very important. It is a good idea to periodically clean the motherboard, the vents leading into the Central Processor unit, and the filters that cover these vents. Accumulations of dirt and vapor pollutants coming into the Central Processor and building up on the motherboard can raise havoc with the circuitry. Inspect these areas periodically for build-up of dirt and any traces of corrosion. The more pollutant-free the environment around the computer, the less you will need to do this. Keep the floor in computer areas vacuumed or swept with an oil mop and the equipment dusted as well. When accumulations are detected, use a can of pressurized air to blow the contaminants out and away from the computer parts.

Sometimes malfunctions result from dirty contact points on interface cards or chips. When operating problems occur with one of the peripherals, or your program starts doing strange things, and you have made certain that the cables are properly connected, the problem might be faulty contacts on the motherboard.

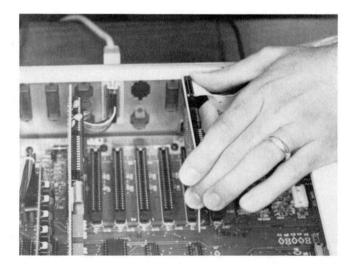

If you are having trouble with your disk drive, check the disk drive interface card, or if the problem is in the printer, the printer interface card, etc. Gently rock the card back and forth while pulling away from the motherboard connector, and remove the card.

Clean the contact points by rubbing a clean pencil eraser carefully along each of them, or use a Q-Tip moistened with isopropyl alcohol. (An abrasive eraser such as those used to correct typing is even better.) Reinstall the card by gently pushing while rocking the card back and forth, until seated.

For the chips, you may need a small screwdriver to begin the removal process. First, insert the screwdriver under one end of the chip and gently pry upwards, then pry the other end. Once loosened, carefully rock the chip back and forth while pulling away from the socket.

Note which direction it faces as you remove it. Unfortunately computer chips can be inserted into their sockets either way, although only one way makes them work and the other way could be disaster. The key is a small notch molded into one end of the plastic chip, and/or a small round dot indented into the plastic near one end. The socket will usually have a corresponding notch or dot in the appropriate end. Look for these identifying indicators when you remove chips to assist you in replacing them the right way.

This is also a good time to mention handling of the chips when they are out of the circuit. Some chips are vulnerable to static electricity. Just the static you pick up from walking across a rug or carpeted area can be sufficient to destroy some chips. When handling them, try to keep your fingers off the little metal contacts. There is really no need to touch them. If the work area is carpeted, try to find a place near a water pipe or metal structural element such as a post or I-beam. Failing this, use the small center screw that holds the wall plate on an electrical outlet. What you want to do is touch this metal before you start working with the chips, and again once-in-a-while while you are performing the work, to "ground" and discharge any static charge you may have picked up.

Clean chip contacts only with alcohol because the eraser could damage the delicate pins. Being certain that the chip is facing the same direction as when removed, reinsert it. If you are performing regular maintenance by cleaning the contacts of your cards and chips, do only one at a time to assure that you reinsert the right card or chip in the correct position and direction as removed.

Another problem that may develop within your Central Processor is the building-up of heat. The room in which your computer is housed may be subject to high heat levels if there is no air conditioning, or if the winter heating system is hard to regulate and runs hot. Add to this environmental temperature excess a computer that has been ON for some time and you have perfect conditions for heat build-up. This excess heat has the potential of causing your computer components to fail.

Some computer chips, particularly CPUs are large-scale devices with hundreds of components layered into their fabrication. These chips use enough current to heat them beyond where a finger can be left on them. Used within the design of the total Central Processor unit, operating in a normal room temperature within the 70°'s F., the chip will operate within its specified tolerable temperature. But excessive heat, such as from operating in a zone of direct sunlight, can easily cause temperatures within the unit to soar, making it likely that normal operating range will be exceeded and the chip will self-destruct.

The foil conductors of some computer circuit boards are extremely close-spaced, with even narrower points where solder connections have been made. It is conceivable that slight expansion caused by excessive heat within the unit could cause a short circuit. Attempts to keep computers cool are the best insurance against failures caused by heat build-up. If the computer has ventilation slots, for example, it is important that they be allowed to do their job. Keep books, papers and materials that you are using away from these vents. In addition, as stated above, keep vents and their filters as well as the motherboard clean. If a computer has a finned or ribbed component such as a separate power supply, locate it where air can freely circulate around it.

When deciding on the areas to place your computer, do not locate it in natural heat conduction areas. Especially avoid placing it near heating vents, radiators, or near windows where prolonged sun exposure can occur, e.g. western or southern exposure.

If you have a computer lab setting with a row or rows of microcomputers on tables in a tropical climate and without air-conditioning, you might consider installing a duct across the back of the table to bring forced air from a common fan or blower at one end of the table. The flexible duct used with clothers dryers, combined with PVC pipe Ts and a fan originally intended to vent a shower or to cool stereo or other electronic equipment will usually permit channeling enough cooling breeze to each computer to keep temperatures well within safe operating limits.

If you have been using the computer for an extended time, and the room is rather hot, heat build-up problems can either be prevented or corrected by simply turning the computer off for a period of time. Of course, a fan on the inside of the computer will greatly cut down on the build-up of heat within your computer. Not all computers have a built-in fan, but accessory covers are available for some models that add a fan for internal cooling, or that at least provide additional vents. Improved internal cooling is especially beneficial as more interface cards are added to your system.

Any or all of these procedures will greatly help to keep heat build-up to a minimum and improve reliability by reducing computer component failures.

Two other external factors that can affect a Central Processor are static electricity and magnetic fields. We have discussed static precautions when handling chips outside the circuit, but the whole Central Processor unit is similarly vulnerable. To guard against the influence of static, try not to place computers in a carpeted area, or if that is not possible, consider using an anti-static mat under the computer.

Magentic fields can be in the form of radio transmissions, as from a CB transmitter or wireless telephone, or actual compass-deflecting fields such as those near large color TV receivers and stereo speakers. The smaller computer monitors either have weaker fields, or are shielded to work near computers.

Finally, among external problem causes are "line spikes," sharp power surges that occur on power lines because of the starting and stopping of motors, lightning somewhere along the line, and instantaneous power interruptions and restorations. A small and inexpensive spike suppressor will usually eliminate problems resulting from these causes.

There is also a category of computer problem that falls into the "if it works, don't fix it" class. The difference here is that for some reason it is not working quite right. Most readers of this book are working in learning institutions, where most of the computer users are at least inexperienced learners. These people can give a computer impossible commands and how the computer will react to these commands is unpredictable, to say the least. Added to these people problems the computer has to deal with a number of the factors discussed above. It has even been suggested that an occasional cosmic particle can permeate the circuitry, producing strange, inexplicable temporary symptoms that defy logical analysis. Overnight or over-the-weekend rest will often allow the unit to normalize itself and the problem will disappear without deliberate service intervention.

Most recently, some computer labs have been afflicted with computer viruses. These deliberate program-initiated malfunctions are most rapidly spread to small computers connected to larger computers through a data network, either by a dedicated data line or by a telephone modem. But the problems are not limited to network-connected computers. A virus can also enter a computer via a contaminated diskette. Once implanted, the virus glitch can continue to work and spread to "clean" disks, even though the contaminated diskette has been removed.

A virus may take many forms, depending upon the programmer's ingenuity and intent. Some are a clever nuisance, while others maliciously destroy stored records or render large sections of memory useless. Unless the virus is programmed to display a message announcing its presence, diagnosis of the presence of a virus can be quite difficult. Any file that does not behave as expected could be suspected of viral contamination, although the probability is greater that it is computer malfunction.

Categorically, it can be stated that computers utilizing hard disk drives are more susceptible to a virus because a number of viruses work through the Disk Operating System (DOS). Viruses are software problems, so any diagnostic procedure designed to differentiate software from hardware problems should lead to some reasonable conclusions.

If the current disk is having trouble, try the backup disk. If that, too, is showing problems, try going back and loading the original program, then the data disk. If the problem persists, try the same sequence of disks on another computer. This should establish whether you have a hardware or a software problem.

Remember that the likelihood of a virus is slight; if you are convinced a virus is loose in the lab, seek the most expert help available. Decontamination may involve a series of steps necessary to isolate and destroy or decontaminate all infected software. Virus Eliminator Disks of increasing sophistication are also on the market and may provide the poison pill needed to clear your lab and return operations to normal.

More consistent problems are localized by service technicians using diagnostic routines that are designed to put a computer through its paces in a logical, analytical sequence. Lacking such diagnostic routines, it is still possible to resort to random substitution. Whole components such as a Central Processor, Disk Drive, Printer, Cassette Unit, may be exchanged if there are several identical computers in an installation. Similarly, chips may be randomly exchanged, one-by-one, between units, or from a small kit of replacement chips. Memory extension cards often have row upon row of the same chip, making it possible to have just one or two spares on hand for repair replacement. Without a good diagnostic routine, the great difficulty of this kind of random service technique is in establishing whether or not a fix has been effected.

With the Central Processor, there is only so much you can do with limited technical knowledge and diagnostic tools. Don't try to be a super serviceperson--you may do even more damage to your equipment. When you have exhausted the diagnostic and service efforts you feel comfortable with, it's time to send the computer to a professional technician for repair.

DISK DRIVES AND DISKETTES

Unlike the Central Processor, a Disk Drive involves both electronic and mechanical components. Introducing mechanical functions into the problem chain can complicate things, although failures of a mechanism are often more amenable to analysis and correction than purely electronic functions, whose workings and non-workings cannot be seen. And since there is a software element involved, problems may not be in the hardware at all, but rather some difficulty with the diskette.

A problem that could be encountered with Disk Drives and Disks is the inability to retrieve data from a disk known to contain previously written data. This could just be a file problem. Start by trying to call up another file known to be on the disk. If the latter one comes up, clearly there is a file problem.

If you cannot even boot the disk, and you know the DOS is not the problem, the question is whether the problem lies in the Disk or the Disk Drive. Some systems will tell you if the problem is with the Disk, or the Disk Drive. One possibility is that the Disk Drive is not connected to the computer. Usually the disk operating system will display a built-in message advising whether or not the Disk Drive is connected. Whether or not there is such a message, check the connectors to be sure they are fitting tightly at both ends. You can also examine the pins inside the connectors to see that they are not bent or damaged.

If everything looks ok but the trouble persists, suspect a damaged cable. If you have another system of the same set of components, exchange the cable. If the exchange remedies the problem and you can now retrieve data from the disk, or if you do not have an exchange cable, open the covers of the cable connectors and look very carefully for a broken or disconnected wire. Sometimes it is necessary to flex the cable to get the loose wire to move away from its connector pin and show where the problem is. A touch of solder is the answer, should you find a loose wire. Troubles along the length of the cable are relatively rare, but sharp kinks and abrasion that

could break a wire are usually visible on the outside sheath of a cable, if you make a detailed examination.

If nothing turns up that would indicate a bad cable, and you have a duplicate system in your installation, try the disk in another disk drive. If the disk still cannot be read, the problem is most likely with the disk. Possible disk problems are dirt on the disk, or that it has been physically damaged. Disks are easily damaged by bending or kinking them during handling. They can be scratched, leaving a minute gap in the magnetic surface. Writing on the labels with ballpoint pens with any pressure can leave an imprint in the disk itself that breaks contact with the read/write head. Accidental overheating can cause the disk to expand, resulting in a conflict as to where the directory thinks the information is located and where it actually is.

The read/write surface of a diskette is magnetic, similar to the material that is used to coat audio and video tape. Magnetic fields can partially erase, or meaninglessly record information on a disk. Problems can result if the disk has been left in contact with, or stored near, a video monitor or anything with an electric motor.

The Disk Drive is probably the most vulnerable component in the computer system when it comes to a need for repair or service. Since the Disk Drive depends upon moving parts to carry out its operations, it is susceptible to mechanical wear and problems related to lubrication or the lack of it.

Just as a tape recorder records or plays back words and music from a magnetic tape, a

Disk Drive reads and writes data to and from a magnetic diskette. The drive has two motors, one to spin the disk and another to move the read/write head(s) (two, if you have a double-sided disk drive) to the appropriate sector on the rotating disk. These heads on the Disk Drive are subject to the same maladies as the heads on an audio or video tape recorder, with the main problem being oxide/foreign-matter build-up. Lost or missing data is often an indication of dirty heads in the Disk Drive.

Much can be done to retard the development of dirty heads. To begin with, keep the environment which surrounds the Disk Drive clear of air pollutants. Reduce dust as much as possible. This includes not placing the Disk Drive near chalkboards and keeping it under a dust cover when not in use. No Smoking signs are appropriate in areas with Disk Drives. Tobacco smoke contains a lot of tar. As the smoke particles get deposited on internal parts of the Disk Drive, the tarry residue inhibits the free movement of the mechanism. As the deposits build up, combining with dust and lubricants, it becomes increasingly difficult for the read/write head to function, putting extra stress on the motor that moves the read/write head. This all means that the Disk Drive is going to require professional service sooner, at the least, or wear out more quickly and have to be replaced, at the worst.

The floppy disk itself can also contribute to dirty heads on your Disk Drive by what it carries with it into your drive. Disks stored or left lying out in the open are especially susceptible to having dust, cigarette smoke, and other air-born pollutans gather on them. Providing convenient closed storage containers for diskettes and training users to keep disks not actually in

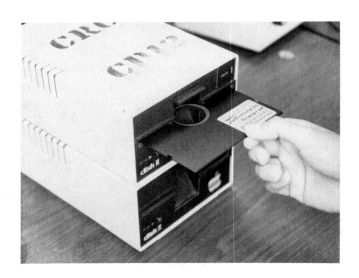

use in the closed storage will reduce this source of Disk Drive problems.

User handling of diskettes should also be monitored, and training given to persons who have poor handling technique. For all the clever design of the disk package, some users will find a way to touch the disk surface through the holes in the jacket, managing to transfer skin oil, soda-pop syrup, and candy smears to the disk. The diskette in turn transfers this mess to the drive heads, and read/write errors predictably ensue.

By removing as much dust and cigarette smoke as possible from the Disk Drive area and taking precautions with handling and storing the disks, you will definitely prolong the cleanliness of the heads and the length of time before cleaning is needed. But in spite of your best efforts, a certain amount of contaminants will slowly accumulate on the heads, particularly from the surface of the disk. The abrasion of the read/write head(s) moving against the turning disk will result in small pieces of the disk's surface being deposited on the head(s). One measure of the quality of a disk is how well the mixture and bonding of the oxide prevents its clogging the read/write head.

It is better to ward off problems before they occur by establishing a regular cleaning schedule. If the Disk Drive operates in a relatively clean environment, cleaning every nine to twelve months should be sufficient. Drives in heavy use (e.g., each hour, approximately six hours a day, at least five days a week) should probably be cleaned every six months. To check for oxide build-up, remove the screws from the bottom of the Disk Drive and slide the cover off

the drive. As you look through the Disk Drive from the front, you will see the spindle that turns the disk and behind it will be the read/write head(s) attached to head support rods. If the head is a yellowish to brownish shade, it needs cleaning. The actual read/write surface is difficult to see unless you have an inspection mirror, but it can just be cleaned on "general principles," since it is likely to have oxide build-up even if the head itself appears clean.

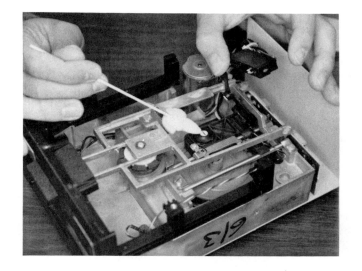

One method of cleaning is to use alcohol. Obtain a cotton swab on a long stick (available through AV or library supply vendors) and isopropyl alcohol. Dip the swab into the alcohol. Remove excess alcohol so that the swab is wet, but not dripping. Wipe the swab over the head(s) lightly, but repeatedly. With the Apple Disk Drive particularly, also wipe the head support rods that carry the head. If this is not done, oxide on the arms will go to the head and you will be back where you started. After wiping the head(s) and arms once, use a fresh swab dipped in alcohol to go over the head(s) and arms again to pick up any residue left from the first wiping. After the alcohol evaporates, the drive is ready for use.

A second method of cleaning disk drives is to use one of the commercially available head cleaning diskettes. While they look like a regular disk, they are in fact a fabric disk enclosed in a jacket. These disks are inserted in the Disk Drive, which is then activated. Some of these disks are used without a cleaning solution and can be abrasive, wearing the head as well as cleaning it. Other types of these disks are premoistened with a cleaning solution, or come with a small bottle of cleaning solution that is used to moisten the fabric disk before inserting the diskette and cleaning the head. If you use a diskette that requires moistening, be careful not to use too much of the solution. Moisten the disk, do not wet it. Since cleaning disks do not clean the head support rods, they are not as effective for the Apple Disk Drive.

Any use of cleaning disks should be done with caution. Use only those that are stated to be non-abrasive and keep cleaning records to assure that cleaning is not overdone. The disk should only be used briefly each time it is used in the drive, and should be disposed of when it becomes dirty. Specific directions are provided with each cleaning disk and these should be followed. Vendors offer a variety of head cleaning kits which include these head cleaning diskettes.

PRINTERS

Before we address some of the more common problems encountered with printers, and their possible causes and solutions, it would be appropriate to emphasize the importance of preventive maintenance. The Printer should not be overlooked in the preventive maintenance schedule just because it seems a little removed from where we see most of the action taking place, i.e. the Computer and Disk Drive. The preservation of cleanliness in the Printer is, if anything, more difficult because particles of paper released by perforation and pin-feed hole drilling of the paper accumulate on mechanical surfaces within the Printer, forming a residue that is ultimately certain to gum up the works.

Schedule your Printer for a good cleaning at least twice a year, more often if you notice dust and debris collecting. First, unplug the printer. Next, use compressed air to remove dust and fine paper deposits that have been accumulating. (Strong vacuum is also very effective, but be careful that it doesn't swallow up the ribbon.) Concentrate especially on the print head area, including the rods along which the print head travels.

Remove any accumulation that tends to pile up, looking almost like a felt washer, at the ends of normal travel on the rods.

Depending upon how much your printer is used, the print head may need to be cleaned more than twice a year. If you use compressed air, make sure that the rest of the computer system is covered so you don't just end up moving the dust from one device to another.

If you notice that some of the deposits still remain after the compressed air or vacuum treatment, apply some isopropyl alcohol to a cloth and use it to finish the job. It is very important, however, that you keep the alcohol away from rubber parts. Also, occasionally you will need to clean the print heads

more directly than with just compressed air.
Isopropyl alcohol on a cloth works very
nicely for cleaning either daisy wheels or
thimble heads. Do check your printer
manual to be certain that the manufacturer
does not prohibit the use of alcohol on the
head.

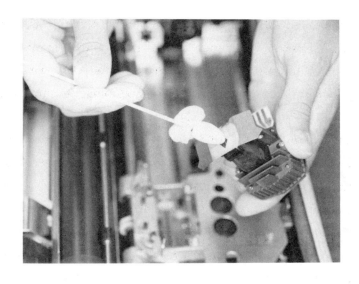

On dot-matrix printers simply remove
the ribbon and print on paper until the ink
build-up is removed. If further cleaning is
needed, use a cotton swab with alcohol on
the end of the print head to clean the print
wires. A special platen cleaner can be purchased to wipe down the platen.

Now let's look at some problems that can be encountered with printers, and their possible
solutions. One problem is that a "print" command is given but nothing happens. The Printer
just does not print at all, or prints for a while and then stops. Assuming that the "print"
command that you just gave is correct, first check to see if the power button is in the ON position and the power light is on. The printer may have become unplugged from the outlet, or
the power cord may not have been firmly plugged in to the Printer. This may seem obvious,
but the obvious is sometimes overlooked until later in the diagnostic process.

Is the "select" or "on-line" button ON? This must be "ON" for the computer to communi-
cate with the printer. If the button will not stay "ON," a lid may not be tight on the Printer.
Next, check the printer cable to see that it is firmly attached to the Printer and to the interface card. Also, check to see that the interface card is properly seated. If everything appears to
be tight, a wire in the cable may be broken. Try moving the cable slightly while executing
a "print" command to see if intermittent printing occurs. If it does, and you have one, switch
the printer cable with one that you know works. If the Printer works fine, the problem is
more than likely in the cable wire itself, a connection solder joint in either end connector, or
a poor connection on one end or the other.

If after all of this checking and switching, you find that the Printer is still not printing,
make sure that nothing is keeping the carriage from moving on its cable. If this is not the
problem, the ribbon may have reached its end (some printers have a built-in sensor for detect-
ing when a ribbon is at its end). This can be very baffling if you know that a new ribbon
has been installed recently and the Printer has had very little use. Try turning the ribbon
advance knob. If nothing happens and you know it is a fairly new ribbon, suspect a jammed
ribbon. Try removing the ribbon and working it back and forth. This might result in unjam-
ming it, or just getting your fingers full of ink. If the latter happens, put a new ribbon on
the printer and try printing again. If upon first turning the ribbon advance knob you find

that more ribbon can be advanced, it may mean that the ribbon advance mechanism is not working properly and needs the attention of a professional technician.

Of course, if you're using new software, do not overlook the possibility that this software has not been configured for your particular printer.

Another printer problem occasionally encountered is the adding of extra letters or leaving out some letters. If the problem just happens once, it may be only a fluctuation in the electrical line. But if the problem persists, check further. It is helpful to isolate whether the problem is in the computer or the printer itself. One way to do this is to run a self-test on your printer. If this is possible with your printer, try it. If you do not get missed or extra letters during the self-test, the problem probably lies in the connecting interfaces or the software.

First, make sure as above that the interface connections are tight. It may be necessary to try another printer cable. If available, try loading a backup copy of the software you are using. If the backup performs properly, the culprit is the original disc or a read error when it was loaded. If the backup does not work either, you apparently have a program problem. If you do not have a backup, try another program.

If the problem is still present, another possible cause might be a bad chip on the interface card. If another card is available, try it in place of the present card.

If the printer did not perform properly during the self-check, the problem lies within the circuitry of the printer. After being sure that the problem is not something simple like a loose daisy wheel, refer the system to a technician.

You may find that your printer begins printing characters that are not complete. If you have a dot matrix printer, this problem could be caused by some print wires that are not working properly. A cleaning could be in order (see above for procedure). If a daisy wheel is being used, cleaning can be done with a cloth wetted with alcohol and a sharp object such as a toothpick or a pin. If cleaning does not help, it may be that you simply have some worn letters on the daisy wheel. A worn daisy wheel will also print some letters dark but others light. Such a wheel should be replaced.

If the incomplete letters are consistently tops or bottoms of letters, then the problem may lie in an improperly seated ribbon. Another cause might be a misaligned print head that is no longer hitting the platen straight on. If this is the problem, see your technician.

You may also find that a printer is beginning to print more and more faintly. This problem is usually traceable to the ribbon. It could be that the ribbon is expended or has become jammed. See above for handling jammed ribbon problems. Other possible causes could be that the ribbon advance mechanism or print hammer is bad. A misaligned print head is another possibility. The latter problems should be turned over to a technician. A less obvious cause could be a dirty or worn daisy wheel as mentioned above, which should be replaced.

A computer communicates with a printer in one of two ways: parallel (eight bits at a time) or serial (one bit at a time). Since some computers have a built-in printer choice, it is important that your printer interface matches what your computer requires.

If you exchange printers between systems in an effort to prove a printer problem, be sure you exchange serial with serial or parallel with parallel, or you may not be proving what you think you are. Many computers, however, give you a choice of either parallel or serial interface ports. In such cases, use the port that is the same as the one used in the original printer.

Parallel is often chosen because it tends to be faster than serial. The serial port on these two-option computers is then free to be used for attaching a modem or a mouse, for example.

III. MATERIALS MAINTENANCE AND REPAIR

VIDEOCASSETTES

Unlike audio cassettes and tape, which can be spliced fairly easily, splicing video tape, in or out of cassettes, is absolutely a matter of last resort. The problem is that it is all but impossible to get the tape ends butted together perfectly enough to prevent some of the adhesive from getting on the heads as they pass over the splice. Such dirt on the heads will degrade the image until it is either cleaned off with a solvent, or until it wears off along the oxide surface of the tape.

It is also not recommended that tape which is badly wrinkled as a result of having been "eaten" by a machine be used again because the creases of the wrinkling can physically damage the heads, and the oxide is usually loosened by the sharp creasing of the plastic base, increasing the possibility of clogging the head gaps.

If you decide to cut out a section of wrinkled tape and try a splice, consider the following:

Because damage often occurs as a result of the internal loading and unloading operation, the bad section may occur at the beginning or end of the tape in the cassette. It may therefore be possible to make a splice before the actual program begins, or after it has ended. It would be great if we could just pull out the clear leader or trailer and attach the tape directly to the appropriate hub, but this would cause trouble because the sensors that control the loading and unloading are photoelectric, and since the tape is opaque, the clear leader and trailer has a definite function in machine control.

The two ends to be joined should be overlapped, held tightly, and cut simultaneously, usually on a slight diagonal. It is important to remember that the video information is read off the outside of the tape, as we look at the cassette, and that it is therefore necessary that the splicing tape be attached to what will be the inside surface when the tape is wound back into the cassette. Working on a white sheet of paper so you can see what you are doing, one person should very carefully butt the two ends together in perfect alignment, with absolutely no gap showing, but no overlap either. A second person then places a piece of approved video splicing tape across the gap and lightly burnishes it in place with a finger. Turn the tape over and check that there is no visible gap. If there is, try the taping procedure again. The splicing tape is usually silver and will really show through any gap that might occur.

When you have what you consider to be a satisfactory splice, burnish the splicing tape to the magnetic tape with the curved part of a fingernail, and very carefully trim the excess splicing tape from either side of the splice. Ever so slightly this cut should remove just a trace of the recording tape too, to assure that the adhesive does not catch on the various tape guides in the recorder/player. Using a Q-Tip or head-cleaning applicator slightly moistened with head cleaner, clean away any finger grease that may be on either side of the tape. Then wind it back into the cassette.

It would be best to immediately order or budget a replacement cassette to avoid probable trouble. If the cassette tape was damaged during machine operation, it would also be a good policy to insist on checking the machine (see the section on VCRs) with special attention to adequate belt tension for Fast Forward and Rewind, and for clean capstan and pinch roller. A slipping drive wheel in the load and unload mechanism may also prevent full-cycle operation. This can cause a machine to wind tape into the mechanism.

If the plastic cassette is damaged, replacement may be possible. VHS cassettes are held together with screws and can be carefully disassembled, the tape turntables removed, tape and all, and inserted in a new cassette case. Because case designs may differ, the old tape should actually be wound onto the new turntables, the new unused tape being discarded. Winding all that tape manually is a painstaking process, but if the result is saving an expensive or irreplaceable tape, it is probably worth it. If the old and new cassettes are of the same brand, the exchange of the complete tape should present no problems. The cassettes are something of a puzzle though, and they need to be taken apart and reassembled very carefully to be sure that the protective door and hub locking mechanisms work properly.

Installations circulating large quantities of video cassettes and concerned about the condition and quality of the tapes should consider the installation of a cassette inspection machine. These machines generally perform a cleaning/polishing function to remove any loose oxide while they speed-check some aspect of the recorded signal and tally a count of problems, thus giving some indication of tape condition. Periodic spot checking by playing a random section of the tape on a VCR would also be advisible. The ease with which an obscene segment could be dubbed into an otherwise innocuous title is a worrisome speculation.

A SIMPLE COMPARATIVE TEST FOR VIRGIN TAPE QUALITY

There are several characteristics that we are concerned about when buying blank cassettes in quantity for institutional use. Price is always a consideration, especially in bid procedures for multi-case lots. The tape should not clog heads. Asking colleagues up and down the road what they are using can be helpfel in this respect, especially if you further ask if they have had trouble with any brand or series of tape. Dropout is usually a strong comparative factor. All tape, audio and video, will have some dropout or variation along the oxide coating. But the less the better, all other things being equal.

Virgin blank tape will produce snow when it is played on a VCR. If the tape is recorded with no incoming camera or tuner signal, as in the "VIDEO" or "CAMERA" position with nothing plugged into the video or camera input, a black raster will be produced when the tape is played back.

The procedure is to record a few minutes with no input signal, rewind the cassette, and then play it back. Sit back and watch the whole screen, counting the number of white flashes that happen randomly anywhere on the screen. These are dropouts. Do this for either 15 or 30 seconds, and write down the number for that particular test sample of tape. It's not a bad practice to Fast Forward the cassette and make the same test somewhere near the middle of the tape too. Then record a tuner or camera picture and play that back, just to see how it looks, and to listen to the sound, on music if possible. Playback should look and sound about as good as the original monitoring of the recording. (Because the picture synchronization is controlling tape speed in a VCR, sound quality may not be as wow-free as the playback of a really good reel-to-reel audio tape recorder.) If there are any noticeable difficulties, write that down. When performed uniformly on a set of test samples, these tests can give you a comparative basis for accepting or rejecting a particular tape during a given purchase process.

These tests should be repeated each time another bid is sent out, because manufacturers may change oxide formulations and tape base plastic, resulting in a change for better or worse at a later date.

COMPACT DISCS AND VIDEODISCS

These discs should be played in their entirety as soon after purchase as possible. The majority of them will play as expected with no problems. A few will have tracking difficulties and will have to be exchanged. Mislabeling has also been encountered and is, of course, grounds to request an exchange. (Gummed labels are not a good idea on compact discs because any "foreign matter" might eventually gum up the player mechanism.)

The larger videodiscs should be removed from the sleeves with the same care and technique careful people used when handling phonodiscs. Particularly, some effort should be made to avoid touching the playing surface. Removal of CDs from the "jewel cases" requires learning a new technique and a peculiar finger coordination to grasp the edges of the disc while applying pressure to the center retaining clip to release its hold on the disc. The CD players are straight-forward and almost impose proper edge handling of the discs. Replacement cases are now being offered by larger record stores and mail order houses.

As with the plastic areas of equipment cases, manufacturers are suggesting cleaning only by wiping with a soft, dry cloth, with strokes from center to edge of the disc. Certainly this should remove bits of lint and dust that might adhere to the disc surface and cause tracking or other reading errors. Holding a disc at an angle to the light will often make problem areas visible.

Light fingerprints that you can see may cause no trouble with the operation of the disc. The actual information is recorded within the laminate of the disc, and the optical system is focused to disregard minimal surface blemishes on the outside plastic surface. Deposits from candy, soft-drinks, etc. may be another matter.

No solvent is ever recommended on plastic surfaces because many of them will cause the plastic to craze, or to loose its shiny transparent surface. This would damage a disc beyond repair. If, however, you have a disc with obvious deposits of some substance, and you believe this to be causing problems in playing the disc, you have nothing to lose. Try a weak solution of dish-washing detergent in barely warm water, applying the liquid with a soft cloth or ball of cotton. Let the liquid do the dissolving and try to avoid unnecessary rubbing. If you are successful, rinse the surface with clear water at about the same temperature and stand the disc on edge to air dry or wipe it dry with a soft cloth, again with motions from center to edge, never with circumferential swirls.

It is virtually impossible to restore the polished surface to damaged plastic. Scratches and/or pits in the surface, or permanent rings caused by standing alcoholic beverage glasses on the disc and using it as an expensive coaster will probably require replacement of the disc.

Returning discs to the protective sleeves or cases is good insurance against damage. Storage on edge is always preferable to stacking discs flat. It is always wise to avoid excessive heat, as from a radiator, heating appliance, or direct sunlight. On an information storage basis, these discs are incredibly dense media. Tiny flaws can have large impact on the amount of information affected. Since you can't fix them, use foresight in handling and storage to assure that they perform as expected.

COMPUTER FLOPPY DISKS

Precautions are about all you can take with computer diskettes. While they are another magnetic medium like videocassettes, they cannot be "fixed." Information storage is very dense, meaning that any dimple or flaw will almost assuredly affect the ability to write on and read from the diskette.

The precautions have been widely publicized, but just for the record and as a reminder, here are some of them:

- Do not fold or bend diskettes.
- Mail only in rigid enough containers to inhibit folding or bending.
- Avoid the use of paperclips to attach anything to the diskette jacket or protective envelope.
- Do not write on diskette jackets or on labels attached to jackets with anything harder than a felt-tipped pen, preferably with permanent ink.
- Do not leave diskettes in the disk drive when not in use.
- Keep diskettes not in use in their protective envelopes, preferably filed in a disk file designed for the purpose.
- Avoid storing diskettes near strong magnetic fields such as those generated by some electronic equipment, or near music-system speakers, most of which contain magnets.
- Don't store diskettes in areas of high heat, as above baseboard heaters or in direct sunlight.

The heavy demand for diskettes has resulted in a competitive market for manufacturers and vendors. Diskettes are offered in a wide range of price and quality, creating a caveat emptor situation. Some things you might want to consider, especially if you are involved in evaluation of disks for a sizable purchase, include the following:

CONSTRUCTION:

Does the jacket look and feel flimsy?

Does the jacket offer sufficient tolerance to permit free rotation of the diskette?

How well are the edges sealed? The jacket is supposed to keep out dirt and a more complete seal will perform this function better.

PERFORMANCE:

(These generally assume that you start with a disk that is not write-protected.)

Run a routine copy program. If the disk is faulty, it may not even boot up.

Try the normal initializing procedure on a diskette under evaluation.

Check evaluation diskette with a Read/Write program of the type used as a Disk Drive diagnostic routine. If the original disk performs well but the evaluation copy does not, something is failing on the copy disk.

Write a long text file and run it with a write/read loop program for a number of hours (typical of the use a disk might be expected to get in your installation). Performance at the end of the test period should be equal to that at the beginning.

IV. APPENDICES

A. TOOLS

The photograph below shows what would constitute a basic set of tools required for most of the procedures described in this book. Primarily, they are tools required to gain access for cleaning and lubrication, for minor adjustments and changing belts. Repair procedures for the class of equipment described in this book often require specialized alignment tools and jigs, small torque wrenches, and electronic test equipment quite beyond the scope of most in-house repair facilities.

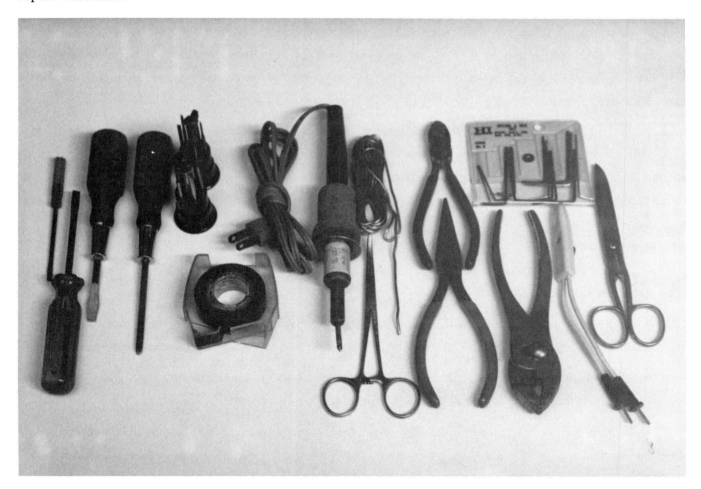

Briefly described, these tools are included:

• Two or more small screwdrivers with blades filed to meet the varied requirements of your equipment. It is always best to try to find a straight blade screwdriver that fits the screw slot snugly because it offers best turning torque and least chance of slipping and damaging the screw slot or adjacent mechanical assemblies. Filing or grinding down the tip of a Phillips cross-slot screwdriver will make it work better with ISO screws. A set of the small jeweller's size blade and cross-slot screwdrivers is useful as equipment becomes increasingly delicate. Outer case screws are usually easy enough to remove, and the inner support structure will take the pressure required to keep the screwdriver in the slot. But do be very careful inside equipment, where screws have often been sealed with a drop of cement and support structures fail to withstand the pressure required to loosen screws. Bending something could affect alignment, and consequently, performance.

• A coil or spool of rosin-core electrical solder, a small soldering iron (25 to 40 watts), and a soldering aid, preferably one with a pointed tip (rather than a brush). The tip of the soldering iron should be small enough to get into the connector pin clusters. If you expect to do a lot of cable repair under the stress of setting up equipment for a program, consider one of the "quick-heat" models. Soldering guns are not recommended for this work because they generate a very strong magnetic field as a result of the high current flowing through the heavy wire tip. This field could cause problems in components like magnetic recording heads and some phonodisc cartridges.

• A roll of plastic electrical insulating tape, while not a tool, should be part of the tool kit. Many brands and widths are available, but Scotch #33 in a 1/2" width is especially recommended for its combination of excellent insulation and good long-term sticking quality. Essential for splicing power cords, it also offers a protective wrap where abrasion has cut through outer insulation, as a safety precaution taped in a layer or two on the inside of a metal cabinet where it is feared wires may be too close when the cabinet is closed, and as a nice finishing touch wrapped just behind the shell of a connector when the cable dimension makes it necessary to leave some shield exposed.

• A 6" pair of diagonal cutting pliers. Indispensable for cutting wires, and even stripping them if you can develop the skill to cut through the insulation only, and then pull it off.

• Some kind of long-nose pliers for reaching into tight places to get hold of things, retrieving dropped nuts and screws, and holding wires while soldering.

• A pair of common pliers. Rarely a true substitute for the proper wrench, they do often go where wrenches won't, and are great for giving that loosening tug to a nut, and holding it while removing or tightening screws with a screwdriver.

• A set of small spline and hex wrenches. These are sold in packs of various sizes in the usual right-angle bent type and in sets of the jeweler's screwdriver type. The right-angle type permits greater turning torque, and often the screws these wrenches work with are very tight. With standard and metric sizes, expect rarely to have the right one and to accumulate a large collection of these over time.

• A neon tester is the most convenient way to check for presence or absence of voltage at an outlet. With one wire held in the hand while inserting the other in an outlet, these testers also quickly show which wire is "hot" (see opening section on Electrical Safety) and provide a final test to prove that equipment cases are not electrically "hot" (with one wire in your hand and the other touching the case, the neon should <u>not</u> light).

• A pair of forceps, preferably with curved jaws. These are great retrieval tools, and a considerable help in coaxing new belts around pulleys that are deep in assemblies, as in VCRs and Carousel projectors.

• Utility knife, electrician's knife, and a dedicated pair of shop scissors. These are primarily used in cable repair. Scissors should not be used to cut wire, but they probably will be, which makes them too dull for normal uses.

The quality and care of tools deserves some attention. Hand tools are the extension of the hands of a skilled crafter. It is fairly safe to say that the person who is sloppy about the storage and condition of tools will be equally messy about the work done with them. Tools should not be loaned. If they come back at all, they usually come back in bad condition.

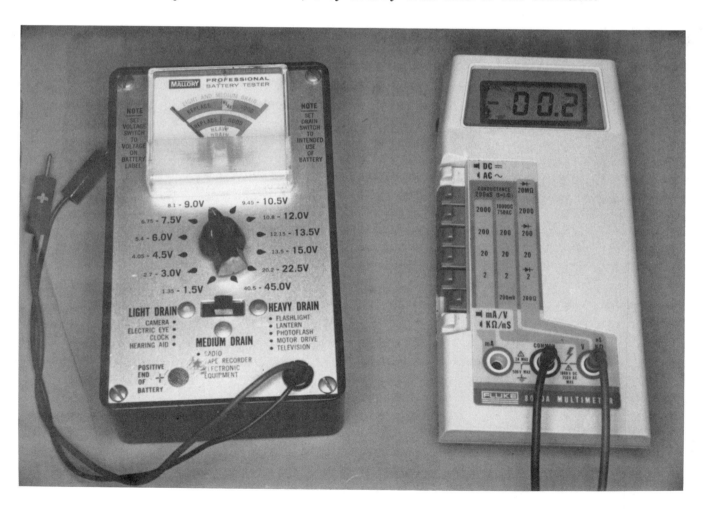

Be careful of low-cost imported tools. Whether imported or domestic, you generally get what you pay for in tool construction and metallurgy. Cheap slot screwdrivers twist in the blade; cheap cross-slot screwdrivers are often so soft they cannot supply sufficient torque without tearing up their own blades; cheap cutting pliers will not hold a cutting edge.

If you expect to do any electrical diagnostic work, some kind of meter is essential. Two that we have found particularly useful are the battery condition meter, on the left, and a digital multimeter. Work at the level described in this book would rarely require a digital meter, but the increasing use of logic circuits with their "yes" and "no" states reflected by a narrow voltage difference might indicate the desirability of a digitally accurate multimeter.

It is important, when using these instruments, to have a sense of significant figures, considering the circuit in question. All that digital accuracy often gives us information that is really not essential or pertinent. For example, when checking a cable for continuity a digital device may indicate a resistance of some tenths of an ohm from one end to the other. This is true. But it is not significant in most cases, and only increasing the physical amount of wire in the cable would reduce it (assuming you are not into superconductivity).

A battery condition meter not only measures battery condition, but does so under an assumed typical load. An old battery might indicate only a few tenths of a volt low on a regular meter, analog or digital, without load. But a battery condition meter adds a load typical of the equipment the battery powers, and in that case the battery may have less than 50 percent of its rated voltage, or held on the battery for a few seconds, may show a rapidly dropping output as it supplies power to the load. If your installation includes any quantity of battery operated equipment, such a meter might be a worthwhile investment. A better investment might be to replace the batteries at intervals determined by the service requirement, whether they need it or not. We found the replacement of nickel-cadmium batteries in older porta-pack VCRs at the beginning of each school year a good guarantee of trouble-free service throughout the ensuing year.

B. CLEANERS AND MAINTENANCE CHEMICALS

Throughout this book, especially in the preventive maintenance sections, reference is made repeatedly to keeping things clean. Because of the great amount of plastic used in the cases and mechanisms discussed, care must be taken to avoid cleaners and chemicals that will dissolve plastics or cause abrasion that would permanently fog the surface of see-through windows.

Most manufacturers now recommend cleaning cases by wiping with a dry or moist soft cloth. If water is used, it should be used sparingly. Persistent dirt might yield to a moistened cloth rubbed on a bar of hand soap. Warm water is usually more effective than cold. Any soap treatment should be followed with wiping with a cloth moistened only with water, as a sort of rinse. It is usually best to follow this with wiping with a dry cloth to reduce rings and spots and to remove the liquid as quickly as possible.

Inside the equipment, water is taboo because of its tendency to cause rust on any ferrous

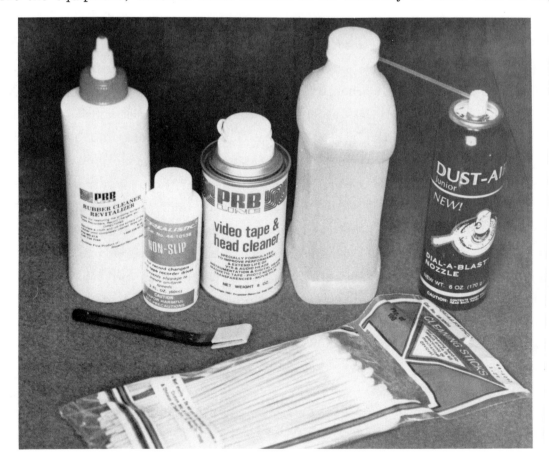

metals that are part of the mechanism. Blowing and brushing are the preferred methods of dislodging accumulated dirt. Because of lubricants and sometimes static cling, the amount of vacuum required to clean mechanical assemblies is appalling, although vacuum can be effective to remove the dirt dislodged by brushing or blasts of compressed air. The most convenient source of compressed air is the aerosol can form, which usually contains a gas to supply the pressure. It is important to use as directed, avoiding tipping the can too far, to be sure air or gas comes out, and not a super-cooling liquid.

Isopropyl alcohol is mentioned repeatedly in the text. Avoid rubbing alcohol because it usually contains some mineral oil. The alcohol is a fine wetting agent that cuts through residues of dirt and lubricant, but evaporates leaving little or no residue of its own.

While many people use alcohol to clean tape heads, it is generally recommended that a cleaner specifically formulated for cleaning video heads be used. Some of the audio tape head cleaners are not recommended for video heads, although the converse is apparently not true and video head cleaner is often used for a wide range of swab-moistening applications.

There are special chamois and plastic foam applicators sold for cleaning tape heads. Cotton swabs are usually too soft to adequately scrub off the oxide, although with enough head cleaner and a lot of rubbing, the oxide does finally come off.

Lubricants have become so highly specialized that we cannot recommend any standard ones for the class of equipment covered in this book. Most equipment has lifetime self-lubricating bearings. Guide rods need to be smooth and polished, but not usually lubricated. Lubricants in things like computer printers are a real nuisance because they cause the fine paper fiber dust to accumulate and slow the mechanism.

Traction, on the other hand, is essential in the belt-driven mechanisms. As the composition rubber belts get older, and as a result of exposure to lubricants and heat, they develop a hard, slick outside surface, even through their tension and compliance is still adequate. This is also true of the rubber tires and paper platens. Treating these by rotating them through a cloth or cotton swab saturated with a no-slip or rubber revitalizing solution will often restore normal operation. It is probably a better preventive maintenance procedure than a corrective one, since the belt in a faulty piece of equipment is usually too far gone to save by mere treatment. This is similarly true of the main drive tire for the loading and unloading mechanism of many video cassette recorders. When these are slipping, replacement of the whole little drive wheel and tire is usually required.

C. SOLDERING

Since the primary concern of this book is preventive maintenance and minor repair, extensive soldering is not anticipated. Still, the replacement of some video heads requires soldering. Most cable repair will necessitate some soldering. Any circuit board that is intermittent when flexed either has a hairline crack that can be bridged with solder, or some soldered connection is not as good as it looks and careful remelting of all solder spots on the board may catch the one that is breaking open.

Soldering is not difficult, if a few simple rules and techniques are followed. Like all skills, it should improve with practice, and some pure practice is recommended before attempting soldering "for real."

The converse of soldering is unsoldering. Repairing a connector often requires clearing the old solder before cutting back the cable and resoldering it into the connector. Disconnecting wires while they are covered with molten solder is not easy, especially since it is imperative not to melt the plastic support or to destroy good components with too much heat. It is as important to become skillful at unsoldering to get things apart as it is to be good at soldering to get them together.

Good soldering requires all of the following:

1. Clean metal, usually scraped with a knife until bright.
2. A good grade of rosin-core solder, either 40/60 or 50/50 (lead and tin). A good grade is not the cheapest you can buy, and it should never be acid-core for any electronics work.
3. A clean soldering iron tip, hot enough to melt the solder instantly when it is touched to the tip.

Before going into the soldering and unsoldering procedures, let's consider some terms and problems. One soldering term is "tinning." We tin the tip of most soldering irons to retard oxidation, which forms a barrier to efficient heat transfer. We also tin wires and terminals, sometimes to hold the strands of wire together and make them easier to manage, sometimes because a pre-tinned wire being soldered to a pre-tinned lug or connector terminal will solder instantly upon application of heat.

"Flux" is another term related to soldering. Rosin is the flux used for electronics soldering and it is manufactured into the center of the solder; hence the expression "rosin-core." There is also an "acid-core" solder, and there are liquid and paste fluxes. These should be left to plumbers, tinners and sheet metal workers. Heat speeds oxidation and oxidized metal does not accept solder. The acid flows over the hot metal, keeping the air off and lightly etching the surface, making a very clean area for bonding.

Acid can also continue to combine with moisture in the air, working over years to eat through thin wires. Acid and lead are basic battery ingredients, and we do not need strange voltages being developed within the circuits. Acid corrosion often approaches a crystalline growth, which if it is conductive to the moisture it attracts, can cause unstable circuit values.

There will be times when soldering just doesn't seem to go right, and you may be sorely tempted to apply some soldering paste found in a shop or relative's tool box. Don't! Only rosin-core solder provides an acceptable flux for electronics work.

The greatest problem in soldering today is the extensive use of plastics, both as cable insulation and as the separating insulator in connectors and components. The key to minimum damage to plastic components is the use of a clean soldering iron tip, hot enough to melt solder instantly upon contact. The most commonly observed bad technique is that of using an iron too cool to melt solder, but plenty hot enough to melt plastics, and leaving it in contact long enough to destroy everything while waiting for enough heat to melt the solder.

Still another problem arises from trying to solder metals that cannot be soldered. Only brass, copper, silver and gold plating can be easily soldered. Ferrous metals (those that will cling to a magnet) cannot be soldered. Aluminum is not easy to solder, and requires special fluxes and solders to get it to solder at all.

SOLDERING

Begin by tinning the soldering iron. Let the iron get hot enough to melt solder, wipe off the tip on a moist viscose sponge or wet rag to remove any accumulated oxide. After long use it is sometimes necessary to use a file to remove oxide and re-shape the tip.

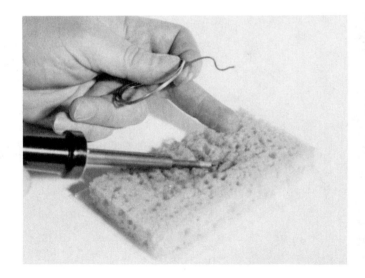

Quickly apply solder to the tip and rotate
the iron to assure that solder flows over
the entire tip surface. Tinning an iron
is an "as needed" procedure and need not
be done if the tip appears clean and will
melt the solder on contact.

A clip aid is often useful, and can be
bought ready-made, or made with an
alligator clip, piece of heavy wire, and
a block of wood.

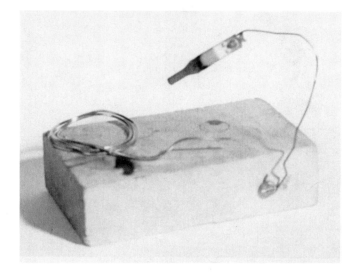

Good soldering requires that the tip of the
iron be hot enough to melt the solder instantly upon contact, and that solder, iron,
and the metal to be soldered come together
simultaneously. The idea is to apply the
heat, get a quick melt that leaves a smooth
solder surface, and get the heat away before it damages any adjacent plastic.

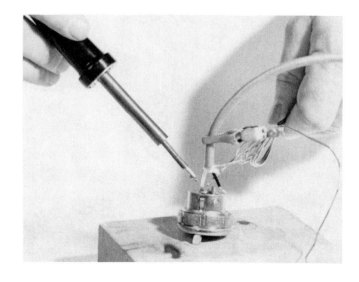

"Tack soldering" is a good technique for temporary, electronically optimum, connections. Solder is melted onto each surface to be joined, tinning the surfaces separately. Then, with a small drop of solder on the iron tip, the two surfaces are touched momentarily with the tip, as the two surfaces being joined are held as motionlessly as possible until the solder cools and sets. A granular, dull appearing surface of the solder is indicative of movement while cooling and should be re-melted.

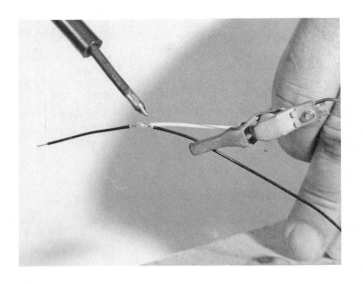

UNSOLDERING

A spring-loaded Solder Sucker is a big help in removing molten solder, especially from circuit boards. The Teflon tip can withstand a lot of heat, and is pressed into the molten solder as the release button is pressed. A piston draws up into the tool, sucking the molten solder away from the lug or circuit board.

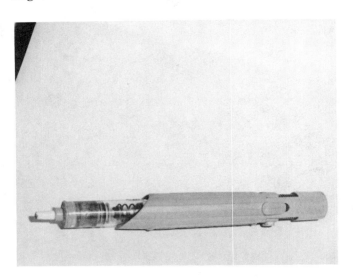

Wires are frequently wrapped around lugs during manufacture of the device, prior to soldering. These can be devilishly difficult to remove while keeping the covering solder molten.

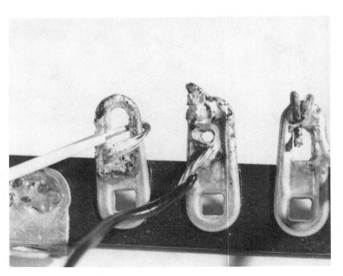

A Soldering Aid with a slit tip can be used to unwrap the wire while the soldering iron tip is held against the lug to keep the solder melted. It is sometimes helpful to cut the wire first, allowing unwrapping from either side of the wrap while the solder is melted.

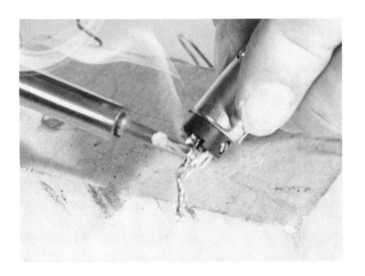

Removing wire from connectors prior to repairing a cable is especially difficult because of the danger of melting the plastic connector insert while digging around in the molten solder.

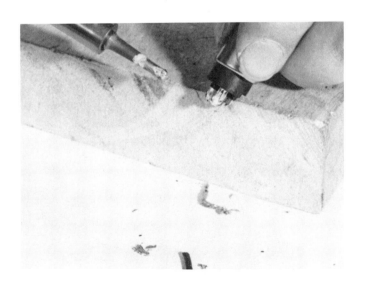

A good technique is to cut the wire as close to the connector lug as possible. Then, while the solder is still molten, rap the connector against a solid surface in an effort to make the solder and wire fly off the lug. It is also possible to be too fastidious in these procedures. Leaving a strand or two of wire to be buried in the resoldering would be preferable to melting all the plastic.

D. CABLES AND CONNECTORS

Cable and connector failure is far more common with institutional use of the types of equipment discussed in this book than it is in home installations. Cable failure is also more common than component failure, which is fortunate since cables are a lot easier to fix, though cable breaks can be deceptive and difficult to diagnose. The constant moving of equipment, with or without connection and disconnection, takes its toll.

The actual wire used in a cable should be appropriate to its purpose. Replacement power cord should be at least as thick (the wire, not just the plastic jacket) as the original cord. Audio can use shielded audio cable, but video and TV RF signals should be connected with co-axial cable of 72- (or 75-) ohm impedance ... not the coax used for CB antenna connections.

Monitor connecting cables may be cut to rather precise, workable lengths. Extending them, or replacing them with longer cables that result in degraded screen image (most notably some kind of vertical line) may require fine-tuning by cutting off an inch at a time until the image is restored to the quality of that obtained with the original cable.

Three-prong power plugs should be replaced with three-prong plugs, preserving the safety feature of the "ground" prong that is connected to the green wire in the cord. If a suitable three-prong plug cannot be found, consider splicing the wires from a short three-wire extension cord to the original power cord wires. Safety standards now require that all power plugs include some sort of insert or construction that prevents the wires from touching the outlet plate.

Examples that follow include installing a power plug, RCA-type audio or video connector, Type-F television RF connector, and a solderable type multi-wire connector typical of those found in computer installations.

POWER PLUGS

Buy replacement plugs that are currently
on the market (rather than use an old
plug salvaged from junked equipment) to
get a product that meets the code. Specifically this means the plug should not
allow wires to touch a metal wall plate.

Assuming the power cord is zip cord, slit
the thin separator plastic with a utility
knife to start separation.

Pulling the two wires apart about 1" back
from the end of the cord.

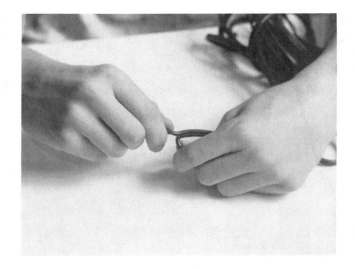

Carefully cut through the insulation all
the way around, cutting no wire strands,
if possible. A dull knife is some help
for this step. When the insulation is cut,
pull it off the wire.

Twist the wire strands together in a clockwise direction (as viewed from the end of
the cord). Twisted the right way, the
strands will pull together under the connecting screw. The wrong way fans them out.

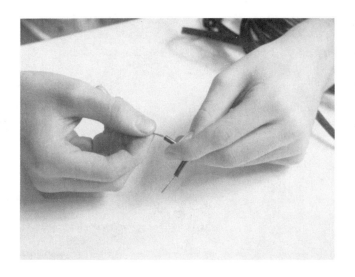

Insert the wire through the plug shell.
You may want to do this before preparing
the wires. If you never forget this step,
you may be too organized for this kind of
work.

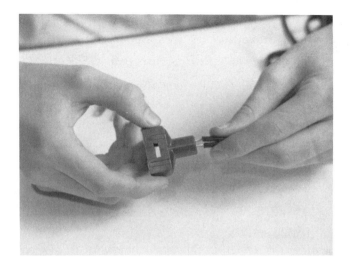

Bend a hook in the twisted wire and hook
it under the connecting screw. Try to
work the tail of the hook under the screw
with a screwdriver.

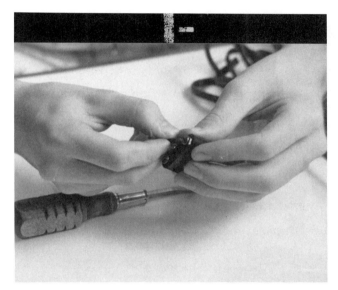

Tighten the screw on the wire, making
every effort to assure that the tail of
the hook stays under the screw. If it
works out, loosen the screw, retwist the
wires, and start over. When you have
finished one wire, do the other one,
fastening it under the opposite screw.

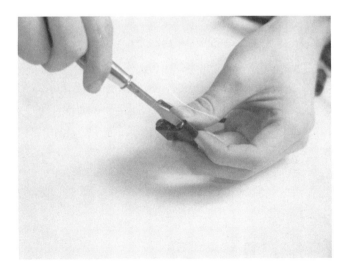

Assemble the plug, watching carefully to
be sure there are no strands of wire either
shorting between the screws, or protruding
from the shell. This completes the replacement.

GROUND-PRONG PLUG

Replacing a ground-prong plug presents
a special problem because no truly small
cord plug of this type has appeared on
the market. This can be a real problem
with equipment that provides a storage
compartment for the power cord. Either
replace the whole cord or cut a molded
plug from a short extension cord ...

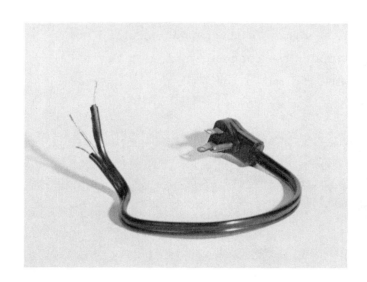

... step-cutting the wires from the plug
and the device, maintaining the color for
color of the insulation through the splice.
It is especially important to be sure the
green ground wire in the center of each
cord be joined in the splice.

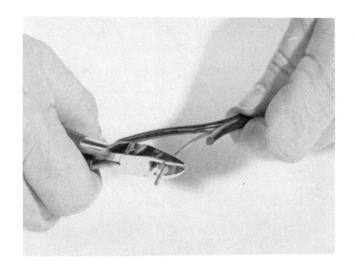

After twisting the strands together in
each wire, twist the pairs to be joined
together. While not essential, it is preferable to solder these wires for perma-
nence and best electrical conductivity.
Bend the wires as shown in the photo,
trying for a conformation that would
prevent any wire touching another
wire.

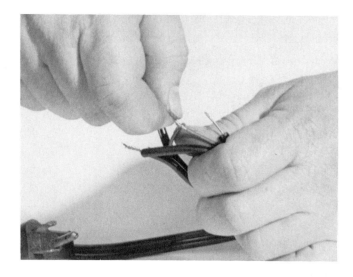

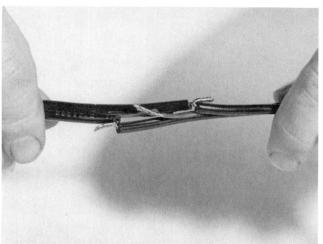

Wrap each wire with enough insulating
tape to fully cover the bare wire joint.

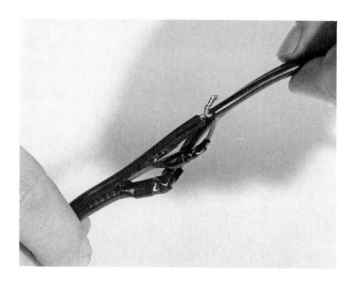

When each wire is fully wrapped, start over with a roll of tape and wrap the whole cord. This wrap should be tight and including the overlap, about two layers thick.

The final job will not be beautiful, but it will provide a serviceable replacement and maintain the integrity of the grounded plug.

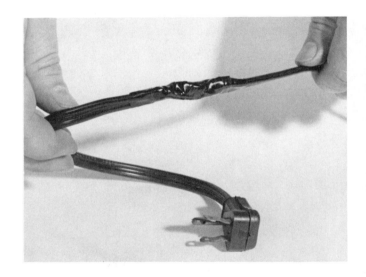

RCA PLUGS

There are two difficulties in doing a satisfactory job of replacing RCA plugs, or fabricating a new cable with them. One is aluminum, either in the shell of the plug or the shield of the cable, and the other is getting a good soldered connection in the center pin. The better quality screw-on shell plug is easier to work with, provides a grip when inserting or removing the plug, and gives a better looking job. If you must use the second type illustrated, try to buy some with bright shells, which are probably not aluminum and can be soldered. If the cable shield is aluminum and will not solder, twist the strands of aluminum wire together (they won't want to stay together either), wrap the wire around the plug shell, and then wrap with insulating tape as tightly as possible. The screw-on shell plug usually has a small lug that can be crimped around the twisted shield wire with a pair of long-nosed pliers. Cut away any excess strand of the shield wire.

Remove the insulation about 2" back from
the end. An easy way to do this is to
make a circumferential cut, then to slit
the jacket from the cut to the end of the
cable. This will allow the jacket to be
peeled away.

Push back on the shield to bunch it up ...

... then use a small tool to unravel the
woven strands of shield. A loss of some
strands of wire is not important, as long
as about half are still coming from the
cable.

Twist the remaining strands of shield together. If the wire will accept solder, now would be a good time to tin the shield to keep the strands from becoming unraveled.

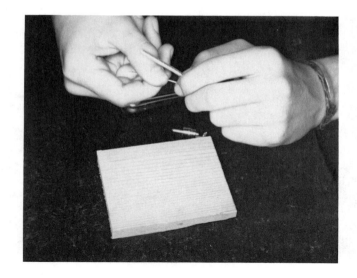

Time for measurement. Push the shell down against the cable jacket, place the plug against the center wire, and note the length of center wire needed to get through the pin. Since any shield wire or foil touching the pin would short the cable, allow a slight pull-back of the cable when soldering the pin.

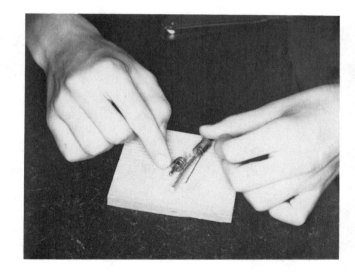

Very carefully cut the insulation of the center wire at the noted point. Try very hard not to even score the center wire itself.

Assuming you removed it, put the shell
back on the cable ...

... and tin the center wire to ease
soldering the wire into the plug pin.

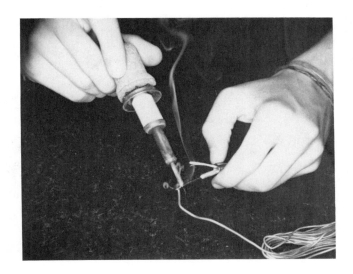

Slide the plug onto the cable, threading
the wire through the pin of the plug.

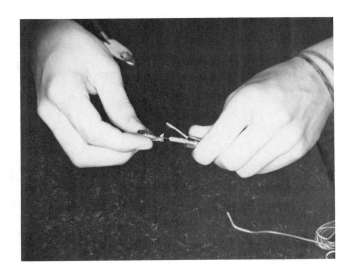

If the wire is a bit long, leave it. It can be cut off later. Solder the wire to the pin. Properly done, there will be a slight roundness of solder at the end of the pin, and when cooled, the wire is secure in the pin when pulled. There should be no excess of solder on the outside of the pin. Too long a wire can be cut off, and small amounts of solder or rosin on the pin can be shaved off with a utility knife after it cools.

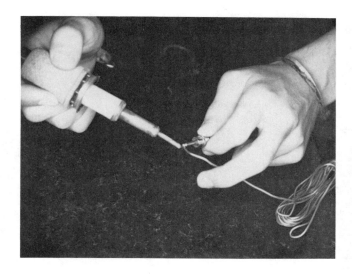

Solder or crimp the shield into the plug lug ...

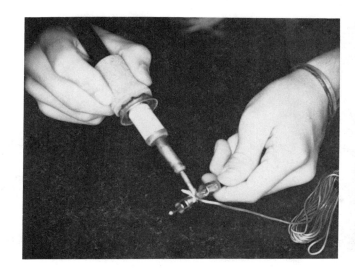

... and cut off any excess strands of shield. Examine the work carefully to be sure no part of the shield metal is touching the center pin.

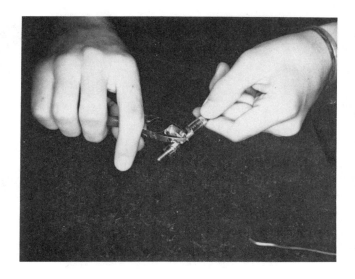

If everything looks good, slide the shell forward and screw it onto the plug.

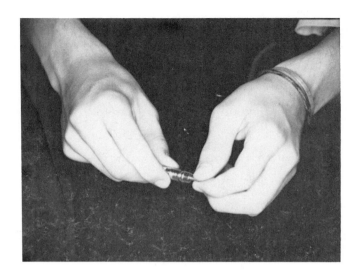

Audio wire is usually small enough to allow the jacket to go into the shell, but coaxial cable will be too large. A tight wrap or two of black insulating tape will complete a very professional looking job.

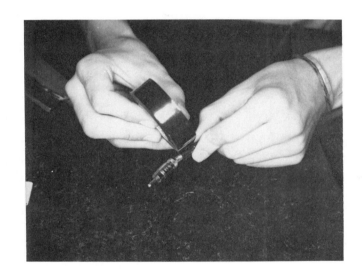

The other kind of RCA plug is easier to do, but these plugs almost necessitate pulling on the cable to disconnect them ... never a good practice. Remove the cable jacket about one inch back from the end and push the shield back.

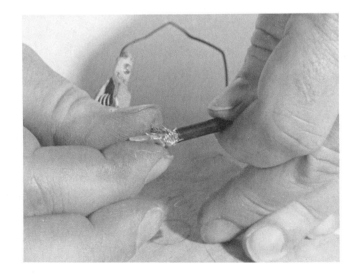

Cut away the insulation from the center
wire and push the wire into the plug,
threading it through the center pin.
Leaving about a 1/8" stub of center
insulation beyond the shield should
assure that only the center wire makes
contact with the pin. Solder the
shield to the plug shell.

Finally, solder the center wire to the
plug pin and cut off any excess wire.
Shave any solder or rosin from the outside of the pin. Properly completed,
the pin should be full of solder, with a
slightly rounded tip.

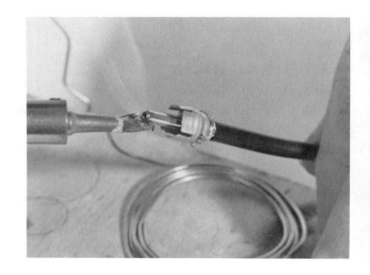

TYPE-F CONNECTORS

Type-F connectors are solderless. Measurement is the essence of doing these
connectors right, but it is not critical.
Use the connector at hand as a measuring
guide.

Cut the outer jacket and shield back about 3/4" from the end. Trim off any strands of shield, but try to leave enough to slightly fan the shield at the end of the jacket.

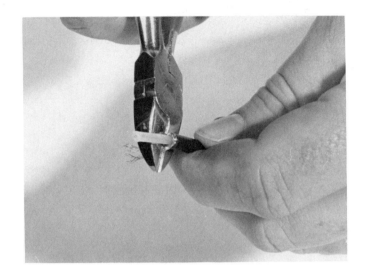

Cut the center insulation about 1/8" to 1/4" from the jacket and shield, and remove the piece of insulation. Be very careful not to cut into the center wire.

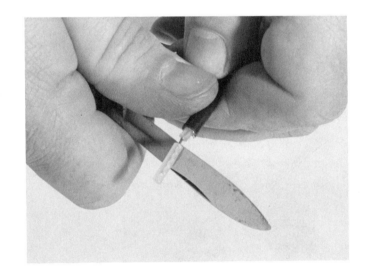

If the connector has a securing ring, put it over the cable first, then push the connector onto the cable. The rear section of the connector has to slide up into the shield. Cable and connectors come in two sizes, so if you do not get a snug fit, you have the wrong combination. It is also possible to get the right combination that is so snug it is virtually impossible to get things together. If this happens, you can slit the jacket about 1". The Type-F connector threads also fit on a headphone jack so the connector can be

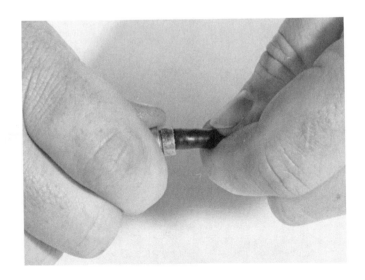

screwed onto one of these (such as the headphone jack on a record player), which will allow a firmer base against which to push the cable on.

Cut the center wire off to protrude about 1/16" from the rim of the connector collar. If the coaxial cable has stranded center wire it will have to be tinned with solder to make it rigid enough to make connection.

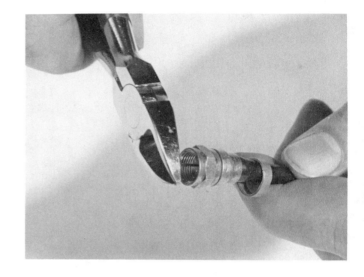

Finally, slide the securing ring over the cable jacket at the connector, and crimp the ring with a pair of pliers or official crimping tool. If there was no securing ring, crimp the back of the connector, but don't overdo it and flatten the whole connector.

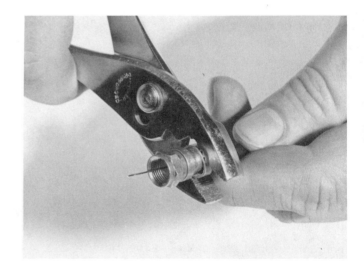

MULTI-WIRE CONNECTOR

Cut and strip away the outer jacket of
the cable, far enough back from the end
to allow the wires to spread across the
connector to be used.

Fan out the separate conductors and strip
just about 1/8" of insulation from the end.
Try to avoid cutting any of the wire
strands.

Tin each bare end of wire. Do not let
too much solder accumulate on the wires
since they will have to be inserted in
the connector pins or lugs.

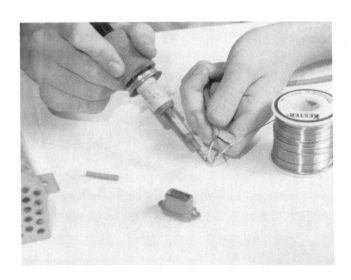

Now tin the pins or lugs of the connector.
Only those which will be wired need be
tinned. CAUTION: When working with
these connectors keep in mind the reversal
of configuration from front to back. A
number 1 pin on the left from the front
will be on the right from the back.

Finally, bring soldering iron tip, connector
pin or lug, and wire together simultaneously,
inserting the wire while the solder is melted
and holding it motionless until the solder
cools. Repeat this for each requisite pin or
lug. It is not uncommon for computers to
use only a small number of the available
terminals. Cross wiring (wired from one
number in the connector on one end to a
different number in the connector on the
other end) is also fairly common. Great
care must be exercised to get each connector wired as specified.

E. AN OVERVIEW OF VARIOUS RECORDING AND REPRODUCING SYSTEMS

It might be informative, if not necessarily useful, to describe the various systems used to record and retrieve information. If nothing else, such a discussion will surely confirm the contention that the new systems are much more complex than the old ones, embracing as they do much of the technology characteristic of the "space age."

TRADITIONAL PHONODISCS

Let's start by looking at monaural and stereophonic analog recording on traditional phonodiscs.

In the monaural system, a stylus cuts a groove into a disc, usually metal, coated with a lacquer or shellac. As the stylus cuts the groove, varying electronic signals corresponding to voice or music move the stylus from right to left and back again, causing the sides of the groove to be modulated with a wavy surface that corresponds to the electronic signals.

By an involved process of electroplating, stripping, and pressing, multiple copies can be made of the "master" recording. These pressings are what we buy as analog disc recordings. If only one record was required, the master itself could be played.

When the disc recording is played, a needle, this time with a rounded point, rides in the groove and converts the modulations of the wavy groove wall into motion, which is then converted back into electrical signals with a transducer, usually a small piezoelectric crystal device in institutional record players (or a magnetic armature and coil arrangement in higher quality units). The electrical signal is amplified and finally converted back into a reproduction of the original sound by the speaker in the record player.

Monaural recordings, still used primarily for some physical education applications like square dance music and exercise routines, usually rotate at 78 revolutions per minute and the reproducing needle radius is in the order of .003 mils. The downward pressure of the needle on the disc may be as high as an ounce or more, or as low as five or six grams, depending upon the age of the equipment and the pick-up cartridge system employed.

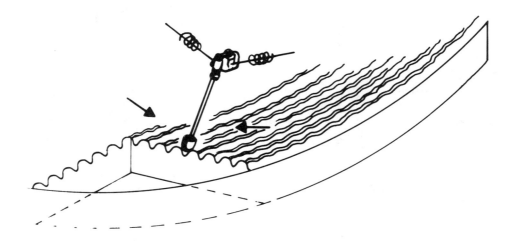

Stereo recordings are a considerable step up in sophistication, although the principle remains the same. Some of the constants get changed. The primary difference between stereo and monaural recording is that, instead of moving the recording stylus from right to left and back, it is moved in two 45° angle movements, resulting in groove walls that have information impressed 90° from one another. Since there is only one actual cutting stylus receiving information for a left and right channel, obviously what actually gets recorded in the groove is a complex vector result of the two signals. It is this composite nature of the groove modulation that limits the channel separation possible from a stereo disc recording.

The needle for a stereo recording has a tip radius of .001 mil. or less. It is fairly important to note this, because obviously this smaller needle point is going to drop to the bottom of a monaural record groove and pick up all kinds of noise from debris collected in the bottom of the groove. It is for this reason that institutional record players generally have some kind of turnover arrangement, allowing either of two size needles to be positioned over a record.

A stereo cartridge into which the needle fits has two pick-up assemblies coupled to the needle, each at 45°, corresponding to the angle of the original cutting head electro-mechanical system. As the record groove moves under the playback needle, two separate signals, generally corresponding to the original left and right information signals, are generated by the play-back head, amplified, and reproduced by a pair of speakers, placed on left and right to simulate the original microphone placement when the recording was made.

The stereo recordings revolve at 45 rpm. or 33-1/3 rpm. The system was developed to take advantage of great advances from the time when the 78 rpm. standard was developed. At the consumer level, the 33-1/3 rpm. microgroove Long Playing records were primarily a way to avoid the interruptions of record changers, particularly in the classical repertoire. Until the advent of the various digital disc systems, the LP stereo records were about the most delicate and precise things large numbers of people ever held in their hands. Obviously, many people lacked the sense of care required to keep these discs in good playing condition.

SOUND MOTION PICTURES

Taking this discussion chronologically, the next development was the sound motion picture; next after monaural recording on discs, that is. In fact, the very earliest "talkies" did indeed just play some records sent along with the film through the theater sound system. If lip-sync is not required, and many educational films use little or none of it, this works fairly well. But feature films since earliest times have depicted intimate human relationships, with closeups of faces. When those actors and actresses speak, lip synchronization is essential or the emotion elicited in the viewer can range from annoyance to downright comic. But to get sound on and off film, a very great obstacle had to be overcome.

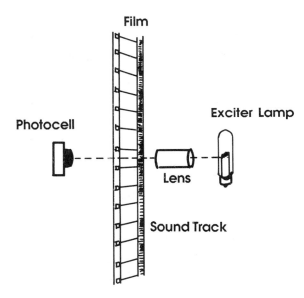

Sound recording demands the greatest possible speed constancy if the playback is to be free of those most obvious of distortions, wow and flutter. But a 16mm. sound motion picture is in fact a rapid display at 24 frames per second of a great succession of very still pictures. If they are not still, the blurring results that is often seen above and/or below the lines of the title and credits.

The problem is resolved by putting a loop above and below the gate through which the beam of projection light passes. Film flows into and out of the gate with the 24 fps. jerky motion of the shuttle mechanism of the projector. In most designs, a smoothly revolving sprocket wheel takes film from the lower loop and feeds it over the sound drum, where the sound is "read" by a narrow horizontal line of light from the exciter lamp, passing through the sound track of the film and on to a photoelectric detector. The sound drum itself is usually at one end of a shaft with a flywheel on the other, which further smooths out the film movement to get the best possible sound. The reading takes place 18 frames down the film from the gate, so the sound is always ahead of the picture on the film. Failure to get the exact number of frames between the optical gate and the sound drum (too large or too small a lower loop) results in the poor lip-sinc often seen in amateur projection.

The sound track is almost always recorded separately from the images we see. On a separate medium, that is. These days it is usually magnetic tape. The clap-stick we have all seen in film-making stories is used to identify each scene, and to get the sound of the stick matched to the frame where it hits the board. The sound track is processed to stay in sync through the scene, and picture and sound are edited separately into an acceptable sequence, using a multiple sprocket-wheel device that keeps picture and sound together.

When editing is complete, picture film and sound tape are sent to a lab, where the necessary lead of the sound is established. As the film is optically printed, the pictures are printed from the edited optical film and the sound is recorded by a light-gate device that converts the electrical impulses from the sound-track tape into a photographically reproducible pattern of modulation. You can actually see this along the edge of the film opposite the sprocket holes.

During projection of the film, a very bright light passes through the projector gate and the film to reproduce the picture on the screen. A much dimmer slit of light reads the sound track at the same time, though 18 frames ahead of the picture, and the film sound-track modulation is converted into electrical impulses which are amplified and played via a speaker into the room where the image is displayed.

These descriptions are reduced to barest essentials for purposes of comparison. They are accurate as far as they go, but they are given with apologies to those who have labored half a century to discover and invent basic methods, and to improve the many processes step by step to bring us to the incredible state of the art we enjoy today. As they say, "It ain't been easy."

MAGNETIC RECORDING

Next in the chronology comes magnetic recording. For the purpose of this discussion we will jump over the historical evolution and move directly to audio cassettes--the most prevalent form in use today.

Magnetic recording marked a radical departure from earlier forms of disc or film recording because no mechanical element was required to transduce the signal. A combination of bias and information signals is mixed and applied magnetically to the tape recording medium. Magnetic domains within the coating of the tape are rearranged or magnetized in accordance with the signal, and unless disturbed by a strong external magnetic field, the domains will return

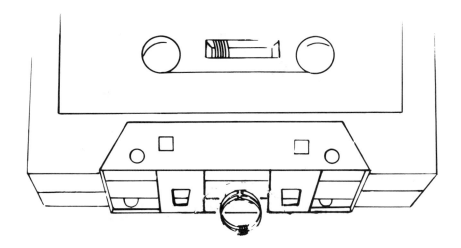

the information signal as a playback if the tape is passed over a head similar to, or the same one, that recorded the signal in the first place. The tape must be recorded and played back at the same speed, 1-7/8 inches per second in the case of standard audio cassettes.

Theoretically it should be possible to record magnetically without channel crosstalk. Practical considerations limit this because the cassette tape itself is only 1/8" wide. At best this would make each track--two in each direction--1/32" wide. But some gap must be left between the tracks, so the actual width of each information carrying track is narrow indeed. The stereo is recorded and played back from the lower tracks as the cassette is in the machine, with the two channels in-line. This results in a very close and tiny head assembly. Channel crosstalk, while markedly better than on a stereo disc recording, is present to some degree.

The very low speed of the tape in these machines makes them vulnerable to problems with wow and flutter. Wow is a long-term perturbation in the mechanical system that would make a cassette of piano music have a twangy sound. Flutter, on the other hand, is a much faster speed irregularity that can be so bad as to make voice seem to have a warble. Generally speaking, the faster a mechanism runs, the less susceptible it is to these mechanical problems.

Unlike the disc and film media, magnetic tape can be erased and re-recorded. In practice this happens whenever the machine is in the RECORD mode, which sends enough bias signal to the erase head to effectively remove any normally previously recorded signal. Under some conditions, where a tape has been over-recorded, the erase function may not completely remove previous signal on a single pass. Since erasure is a matter of realigning the domains to a neutral position, turning the volume all the way down and letting the tape pass through the machine in record mode prior to re-recording will usually clear the tape, since it gives a double erasure ... once on the pure erase pass and again during recording.

If audio cassettes could find the market they did, what could be done to make consumer recording of video possible? Networks had used huge 4-head quadrature-scan video tape

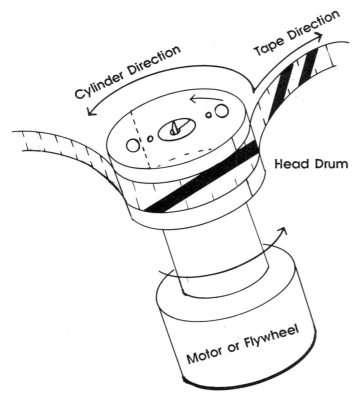

recorders with 2" wide tape for years, so clearly it was possible to record and reproduce video. But machine and tape costs made this system prohibitive in the home market.

The primary problem was how to achieve a recording speed that was mind-bogglingly beyond anything audio requires because of the very high frequency response range required to reproduce a picture with a scale that extended from black through all the shades of gray, to white. This was done by spinning two heads exactly 180° apart against the tape at 1800 revolutions per minute. After a brief period of use of 1" tape, 1/2 inch became the standard, moving at a relatively slow 7.5 inches per second. A helical guide slants the tape, causing the spinning video heads to "write" the picture in a series of diagonal

lines across the center of the tape. Tracks near each edge of the tape, with stationary heads recording and playing back at the lineal speed of the tape, carry the audio on one edge and the sync and other process signals near the other edge.

Unlike the film medium, which stops 24 times per second to project a still

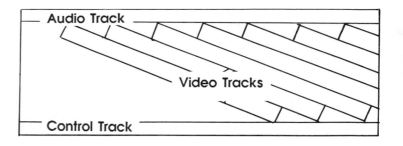

image, the videotape moves constantly. But there is the equivalent of a stop and start as each head comes into contact with the tape, moves down the helical path, and leaves contact with the tape, at the exact instant that the other head touches the tape. It is this scanning motion that allows stop motion viewing of a single video frame. The tape is stopped lineally, but the heads are alternately scanning the same frame line, thus keeping that picture on the screen.

There are a few things that might be pointed out about the numbers in the paragraph preceding the last one. 7.5 ips. was the common high speed of reel-to-reel audio tape recorders before the audio cassette became the widely used format. The parameters and parts for building a reel-to-reel machine were thus known and available in the industry. The figure 1800 is a multiple of 60, and 60 cycles per second is the power line frequency in the United States.

A number of constants used in U.S. television are multiples or sub-multiples of 60. An example is the 1/15 second minimum (1/8 second recommended) for taking pictures of the picture on a television receiver or monitor. In 1/15 second, two alternate line scansions take place, making one complete picture.

Prior to introduction of the videocassette machines that seem to be everywhere today, the black-and-white reel-to-reel 1/2" format videotape recorder was the standard. Common in education and industry, they never really made much impact on the home market. Price stayed fairly comparable to a good reel-to-reel audio tape recorder, largely because new circuitry and component techniques offset the greater complexity of the video machines.

Motivated by the huge market they knew to exist, and some changes in the industry, two groups of engineers in Japan set out independently to develop a convenient videocassette recorder that could be marketed at a popular price, using tape that was easily affordable. There was a 3/4" cassette tape machine (still recommended for initial recording and editing) to use as a model.

At Sony the new machine developed was the Betamax. Using 1/2" tape in a small cassette, it reached the market first and offered outstanding playback quality of color off-air recording. The drawback was running time, initially limited to one hour, as were the earlier reel-to-reel video tape machines.

In a small subsidiary of Matsushita, the other engineers held to a more consumer-oriented specification: an equally convenient 1/2" cassette format, but with a minimum running time of 2 hours, enough to record most feature films. Known as the Video Home System, or VHS, it dominates the video recording market today.

Both Beta and VHS use 1/2" tape enclosed in a plastic cassette. The formats are not compatible, and the threading systems are different, but they both use helical scan recording. Rather than talk about speed, we should turn our attention to running time. In the VHS format there is one interesting difference from what we knew in the old audio reel-to-reel recording. An audio tape 1200 ft. long would play 30 minutes on one half when running at 7.5 ips., 60 minutes at 3-3/4 ips., or 2 hours at 1-7/8 ips. In the VHS system, a T-120 cassette will run 2 hours at the standard play speed, 4 hours in long play, and 6 hours in super or extended long play. Note that the slowest speed does not double the middle one, but rather adds the first and second.

There are also some odd problems of tape cost related to the mass market. Even though you may be going to build a library of titles with 20- or 30-minute running times, it may be more economical to put the programs on T-120 cassettes because they are so common that they are often offered at very low sale prices. The shorter cassettes are surprisingly almost invariably more expensive. Clearly the amount of tape is not the major marketing factor.

DIGITAL RECORDING

Throughout the previous discussion we have been talking about what is known as analog recording. While necessary conversions of energy are performed from mechanical to acoustical to electronic, or from light to electronic and back again, no encoding takes place. At any point in an analog circuit the signal can be considered accessible by appropriate detection instruments for analysis and diagnosis. And those instruments are usually just extensions of similar analog technology, with circuits similar to those in the device itself.

Computers are digital devices. All information is "processed" in digital form. Data cassettes and diskettes store long strings of digitally encoded information. Special encoding and decoding circuitry is part of the electronic path, and program information may be required to instruct the computer or other device what to do with the digital data. An analog cassette tape recorder cannot be used to record and play back data unless a special interface is provided for conversion of the data to something the analog recorder can handle, such as a tone code.

Digital transmissions are generally less vulnerable to interference, as via satellites or telephone lines. Digital information exchange will break down if interrupted, but is less likely to suffer from static and various forms of distortion. Digital technology has much to do with the incredibly fine picture and sound segments from all over the world that occur in network news telecasts.

Before getting into the digital systems, this would be a good time to get one other seemingly extraneous consideration out of the way. The consideration is a lineal medium like tape vs. a disc or diskette medium. The issue is indexing. Anyone who has ever tried to find a specific segment of music or picture on a tape knows the problem of finding just that place to start playing for precisely what they want to hear or show. And not only is it hard to find, but it takes inordinate time for the tape to spool forward and backward during the search.

Disc media are much quicker for indexing. The slowest is trying to drop a tone arm on a record to play a selected band, and that's a lot faster than trying to spool to the third or fourth selection on a tape. Computers have frequent need to access stored data, and lineal media would take forever in an application where time is money and the effort to improve computing speed is never ending. Cartoonists still like to depict computers with reels of tape at the top of a series of tall tape drive cabinets, but the real world has moved on to a low cabinet or cabinets full of stacks of spinning discs. (Cartoonists are not a reliable source. Many still depict schools of the ding-dong bell in the cupola era.)

So if the application is something like showing a 109-minute feature film, a long piece of tape in a cassette is fine. But if accessing data and storing it periodically is required, or if we desire to change the program sequence, as is easily done on Compact Discs, we are much better off with a disc medium.

Digital information is always encoded into digital form, a series of "yes" and "no" bits, strung together into words called bytes. Bytes usually consist of 8 or 16 bits, and the length

of byte a computer can process has a lot to do with the speed of the computer. Typing characters on a computer keyboard encodes the letters and numbers into digital form immediately.
The encoding of analog information such as spoken words, music, or pictures involves a process
called "sampling."

When the information is encoded for transmission or recording in a digital system, the
analog signals are sampled at a rapid rate and the samples are encoded into a digital byte corresponding to the analog signal, according to some predetermined code that is built into the
technology.

When the encoded signal is received or played, the encoded digital information is decoded
by the circuitry and converted into some form that can be interpreted by people, such as type,
pictures, or music. While not all digital languages are the same, as long as the information
stays in digital form it is relatively easy to interface quite complex devices and get them to
work together. The case of this of immediate interest is coupling a microcomputer to a videodisc player to create an interactive teaching/learning device.

During development of the home type VCR, there also emerged a digital disc technology,
marketed under various trade names. Actually, there were two systems, but the technique
involving a needle on the disc has yielded to that using only a laser beam to scan the recorded
surface, which is actually imbedded within the plastic laminates of the disc. The disc spins
at 1800 rpm. (there it is again) and contains about 54,000 frames of video information. A disc
can be flipped, to play either side, making it possible to put a feature film on one disc. That
is probably the market for which the disc was developed, because initially it was considerably
cheaper to manufacture duplicates of these discs than it was to produce copies of lineal videotapes, each of which had to be recorded from one end to the other. Lower prices have created
unanticipated demand for pre-recorded videocassettes, and techniques have been developed to
produce mass quantities at ever more competitive prices.

But the education and training market saw in the coupling of a videodisc, with its incredible 54,000 full-frame picture capacity and ability to display single frames or break away into
motion as needed, plus rapid indexing upon interactive command, coupling it with a microcomputer
to develop the most powerful small teaching/learning machine to date. Not limited to computer
graphics, but able to use actual live-action pictures, combined with the ability to respond to
each learner's needs by branching according to the learner's responses, this combination was
an educational technologist's dream come true. Even the price is not prohibitive for those motivated to embrace this technology. The biggest problem is lack of appropriate disc software
and the very high cost of developing it. Only those teaching areas that can achieve a relatively
mass market can hope to recover development costs and become profitable or cost-effective.

While not identical, video and audio disc recording systems are similar, and at this writing
one manufacturer has announced a single player capable of playing all present formats. This is
a very complex technology compared to what we have discussed before, being a sort of culmination

of most of the techniques developed to overcome design problems in the earlier recording and reproduction systems.

The metallic color (silver or gold) of the disc systems is from a reflecting surface or mirror at the center of the laminated plastic disc. The reflecting surface is deposited on the plastic after the spiral pattern of "pits" and space between the pits has been pressed into the plastic layer that carries the information. This information includes not only the signal or data to be reproduced, but a lot of other instruction information used by the disc player to control speed, to maintain tracking, and to index the selections.

A laser beam is focused through the protective plastic surface of the disc onto the pit and space level of the disc laminate. This is the reason minor surface scratches, etc. do not materially affect playback ... they are so out-of-focus as to be all but invisible to the laser beam reading system.

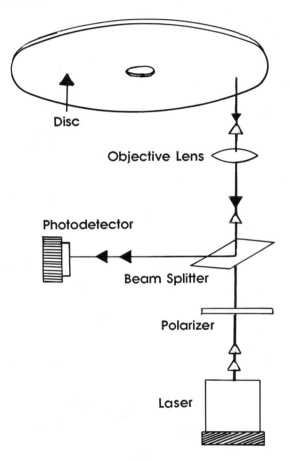

At this point the reading system is somewhat analogous to the optical sound on film of the motion picture projector. The projector doesn't use laser light, and the light passes through the film onto the photoelectric detector, whereas in the disc system the light is reflected back along the optical path, split off, and converted back to an electronic signal by the photoelectric detector. The sound track was in analog form, whereas the pattern of pits and space is a digitally encoded one. Still, the concept of using a shaped and focused beam of light to read an optically recorded piece of information is the same.

Because of the high band of frequencies required by video information, the videodisc spins at 1800 rpm. At that speed the laser is covering pits at a prodigious rate, and in order to have any appreciable playing time, the discs are 12" in diameter, or more recently, some are only 5", with proportionally shorter playing time. The "singles" of the video business.

Audio Compact Discs do not require such high speed to yield a full range of sound. These discs rotate at a variable speed from about 500 rpm. near the center down to about 200 rpm. at the outside (the CDs are played from the inside outward). This is done to overcome an old problem with analog disc recordings called "inner band distortion." As a record rotates at constant speed, the amount of linear groove moving out from under the needle gets less and less as the center is approached and the circumference of the groove gets smaller. But in most cases with Western music, the sound gets louder and louder near the climax. So just when we needed the most groove linearity to handle the greater dynamics, we had the least. A CD changes speed to keep the laser reading rate constant, regardless of its position under the disc. Thus "inner band distortion" or any equivalent is eliminated. Inside to outside tracking also attacks this problem. The CD is 4.7" in diameter.

The CD is not likely to become a classroom record in the near future because of the cost of the discs and the playing equipment. But the indexing advantage would be a clear asset in music appreciation courses and the ability to select constant repeat of a band has possible application for dance classes, for ear training, or for music or spoken word memorization. Since the laser beam imposes no wear on the recorded surface, a CD could theoretically repeat a selection forever.

Also among the digital recording media is the older computer cassette, the digitally recorded floppy diskette, and small hard disk drives that give personal computers fantastic memory power and enable them to carry out some truly impressive business functions like film booking and recordkeeping involving thousands of bits that have to be carried for months.

All of these are basic magnetic recording systems. What makes them unique is the speed of indexing, except, of course, the linear cassette tape.

When the disk drive light comes on, the drive motor is running at about 360 rpm. Controlled by commands from the associated microcomputer, the head (or heads on double-sided disk drives) is/are moved by a second motor (a stepper motor) to follow its orders by getting location instructions from the disk directory and going to a particular track and sector to write or read data. The tracks are concentric circles and the sectors are like slices of pie that subdivide and cut across the tracks. These tracks and sectors are electronically laid down during the initializing of the disk.

The head(s) read and write to and from the disk by coming in contact with the disk in the elongated oval open area of the disk cover. Aside from computer control and digital circuitry, the devices operate by standard magnetic recording and playback principles.

At the time of this writing, the U.S. is on the threshold of the introduction of Digital Audio Tape recorders. Actually, digital audio tape recording has been going on for several years, but the machines have been priced for professional recordists. At issue at this time is a popular format Digital Audio Tape machine, a machine that according to all reports would be capable of copying a Compact Disc with a quality indistinguishable from the original. The hangup is not the technology or the price, but the copyright considerations. Limited copy protection devices have been developed, some to be incorporated in the content of CDs, and others that would have to be wired into all digital cassette machines sold in the U.S.

This kind of social/legal problem is not new to U.S. technology. There are better television systems than the rather coarse 525-line system used in the U.S. But first we were caught by an insistence that any color system adopted should be compatible with all the black and white receivers then in use. Now improvement is forbidden because it would make obsolete all receivers and a lot of the peripherals now in use. Certainly this is a valid social consideration, but it seems a shame to have to settle for less than the best possible because of the earlier adoption of what is now about a 35 year old state-of-the-art.

Digital technology is the wave of the present and future. It is hard to imagine the implication, but those on the cutting edge of these developmental technologies assure us, "You ain't seen nothing yet." Though critical acclaim has not been unanimous, if the vote of the market-place for Compact Discs is any indication, most people would agree that digital has made a quantum leap for the ear.

FURTHER SOURCES OF INFORMATION

PUBLICATIONS

Davidson, Homer L. Beginner's Guide to TV Repair, 3rd edition (Blue Ridge Summit, PA: TAB Books, Inc., 1985).
Going deeper into TV-receiver repair than we think readers of this book should because of safety considerations, the Beginner's Guide would be helpful to a student aspiring to get into TV repair commercially.

Goodman, Robert L. Maintaining and Repairing Videocassette Recorders (Blue Ridge Summit, PA: TAB Books, Inc., 1983).
Slightly dated by now, this book tells quite a lot about how these machines operate, including detailed circuit analysis. Service information is included.

McComb, Gordon, and John Cook. Compact Disc Player Maintenance and Repair (Blue Ridge Summit, PA: TAB Books, Inc., 1987).
This book includes some really down-to-earth suggestions and fine illustrations that should be helpful for maintaining Compact Disc players at the institutional level.

Nayak, P. Ranganath, and John M. Ketteringham. "JVC and the VCR Miracle: You Should Be Very Polite and Gentle," in Breakthroughs! (New York: Rawson Associates, 1986), pp. 23-49.
A discussion of the development of the VHS format.

Pohlmann, Ken C. Principles of Digital Audio (Indianapolis, IN: Howard W. Sams & Co., 1985).
This book gives an exhaustive and very technical explanation of digital encoding, with a non-specific description of how CD players operate.

Schroeder, Don, and Gary Lare. Audiovisual Equipment and Materials (Metuchen, NJ: Scarecrow Press, Inc., 1979).
Although an older publication, this book still stands as a basic source of strategies and procedures for maintaining and repairing many of the types of media equipment found in today's institutional settings.

The following books are good all-purpose guides to preventive maintenance and general repair of microcomputers:

Atwater, Dorothea. First Aid for Your Apple IIe (New York: Ballantine Books, 1985).

Makower, Joel, and Edward Murray. Everybody's Computer Fix-It Book (Garden City, NY: Doubleday, 1985).

Schwieder, Pete H. How to Repair and Maintain Your Own IBM PC/XT (Carson City, NV: Personal Systems Publications, 1984).

Stephenson, John, and Bob Cahill. How to Maintain and Service Your Small Computer (Indianapolis, IN: Howard W. Sams and Co., Inc., 1983).

VENDORS

Demco
Box 7488
Madison, WI 53707
1-800-356-1200
    Demco sells video and audio head cleaning devices, compressed air, tape demagnetizers, audio and video cables, record and compact disc cleaners, small vacuums for cleaning equipment, disk-drive cleaning devices, CRT screen cleaners, printer-platen cleaning kits, printer cleaning kits, computer-keyboard cleaning solvents.

The Highsmith Co., Inc.
One Mile East on Highway 106
Fort Atkinson, WI 53538-0800
1-800-558-2110
    Highsmith is a good source for audiovisual and computer furniture; projector maintenance kits; tools; video-, audio-, and disk-drive cleaning kits; disk-drive analyzers; computer fans; surge protectors; and dust covers.

Prime Electronics Co. (formerly Projector Recorder Belt Corp.)
P.O. Box 28
Whitewater, WI 53190
    This company is a source for replacement heads for VCRs, connectors, splitters, tools, test instruments, and belts. Write and request their complete catalog.

Valiant
Universal Video
195 Bonhomme Street
Box 488
Hackensack, NJ 07602
1-800-631-0867
    Valiant markets a wide array of audio and video cables, adapters, and connectors, as well as many kinds of preventive maintenance supplies and service aids, including head cleaners, compressed air, optical cleaners, lubricants, cleaning swabs, and various types of tool sets.

INDEX

ABOUT THE AUTHORS

DON SCHROEDER was associate director of resource planning and development with primary media responsibilities in the Cincinnati Public Schools. He is now retired. He began his involvement with instructional media as a classroom projectionist in the St. Louis Public Schools, gained invaluable training and experience as a U.S. Navy aviation electronicsman, taught at the high school level, and completed a master's degree in Education (Instructional Technology) at Syracuse University. He has frequently taught the basic audiovisual media course at colleges and universities in the Cincinnati area.

GARY LARE holds the position of Director, Curriculum Resources Center and associate professor of education at the University of Cincinnati. Dr. Lare has taught in the public schools and has been involved in educational media, in both teaching and administrative capacities, at various universities. He holds both a master's and a Ph.D. degree in educational media from Kent State University.